Studies in Computational Intelligence

Volume 501

Series Editor

J. Kacprzyk, Warsaw, Poland

For further volumes:
http://www.springer.com/series/7092

Bo Liu · Georges Gielen
Francisco V. Fernández

Automated Design of Analog and High-frequency Circuits

A Computational Intelligence Approach

 Springer

Bo Liu
Department of Computing
Glyndwr University
Wrexham, Wales
UK

Francisco V. Fernández
IMSE-CNM
Universidad de Sevilla and CSIC
Sevilla
Spain

Georges Gielen
Department of Elektrotechniek
ESAT-MICAS
Katholieke Universiteit Leuven
Leuven
Belgium

ISSN 1860-949X   ISSN 1860-9503 (electronic)
ISBN 978-3-642-44645-0   ISBN 978-3-642-39162-0 (eBook)
DOI 10.1007/978-3-642-39162-0
Springer Heidelberg New York Dordrecht London

Printed on acid-free paper

Springer is part of Springer Science+Business Media (www.springer.com)

Preface

Computational intelligence techniques are becoming more and more important for automated problem solving nowadays. Due to the growing complexity of industrial applications and the increasingly tight time-to-market requirements, the time available for thorough problem analysis and development of tailored solution methods is decreasing. There is no doubt that this trend will continue in the foreseeable future. Hence, it is not surprising that robust and general automated problem solving methods with satisfactory performance are needed.

Some major problems that highlight the weakness of current computational intelligence techniques are appearing because of the increasing complexity of real-world systems:

- Long computational time for candidate evaluations: due to the increasing number of equations to be solved in real-world problems, the evaluation of candidate solutions may become computationally expensive.
- Large uncertainty: the simulations or physical experimental results may be very inaccurate because the human-designed model can only catch the most critical parts of the system.
- High dimensionality: because of the increasing complexity, many currently good human-designed simplified models may no longer be useful, and, hence, the analysis based on these models does not work. Therefore, full models with a large number of decision variables may be encountered in many real-world applications.

From the points above, it can be concluded that new methods with the ability to efficiently solve the problems, methods that can bear large uncertainty and methods that can handle large-scale problems, while at the same time providing high quality solutions, will be useful in the foreseeable future. The purpose of this book is to discuss these problems and to introduce state-of-the-art solution methods for them, which tries to open up fertile ground for further research.

Instead of using many kinds of real-world application problems from various fields, this book concentrates on a single but challenging application area, analog and high-frequency integrated circuit design automation. Since this decade, computational intelligence techniques are becoming more and more important in the electronic design automation (EDA) research area and are applied to many

EDA tools. EDA research is also stimulating the development of new computational intelligence techniques. For example, when searching "robust optimization" or "variation-aware design optimization", it can be found that a large number of research papers are from the EDA field. Moreover, many difficult problems from the EDA area are also cutting-edge problems for intelligent algorithm research.

Therefore, this book: "Automated Design of Analog and High-frequency Circuits: A Computational Intelligence Approach", is intended for researchers and engineers in both the computational intelligence area and the electronic design automation area.

For the computational intelligence researchers, this book covers evolutionary algorithms for single and multi-objective optimization, hybrid methods, constraint handling, fuzzy constraint handling, uncertain optimization, regression using machine learning methods, and computationally expensive optimization. Surrogate model assisted evolutionary algorithm for computationally expensive optimization problems is one of the main topics of this book. For robust optimization in uncertain environments and fuzzy constrained optimization, the state-of-the-art is reviewed; some promising solution methods are introduced elaborately, which complements the available literature. Evolutionary computation spreads throughout this book, but it is not our purpose to elaborate this specific research area, since numerous books and reports are available. Instead, we cover fundamentals, a general overview of the state-of-the-art related to the types of problems for the applications considered, and popular solution methods. In Chaps. 1, 2 and introductory sections of Chaps. 5 and 7, we try to make the beginners to catch the main ideas more easily and then provide a global picture for a specific topic for the use of further research and application. Professional computational intelligence researchers can escape the above mentioned contents.

For the electronic design automation researchers, this book tries to provide a tutorial on how to develop specific EDA methods based on advanced computational intelligence techniques. In many papers and books in this area, computation intelligence algorithms are often used as tools without deep analysis. This book, on the other hand, pays much attention to the computational intelligence techniques themselves. General concepts, details and practical algorithms are provided. The broad range of computational intelligence and complex mathematical derivations are introduced but are not described in detail. Instead, we put much effort on the general picture and the state-of-art techniques, as well as the method to use them in their EDA related tasks. The authors believe that EDA researchers can save much time on performing "data mining" from the computational intelligence literature to solve challenging problems at hand, and even develop their own methods with the help of this book. In addition, to the best of our knowledge, this is the first book covering systematic high-frequency integrated circuit design automation.

The concepts, techniques and methods introduced in this book are not limited to the EDA field. The properties and challenges from the real-world EDA problems are extracted. Researchers from other fields can also benefit from this book by using the practical real-world problems in this book as examples.

Chapter 1 provides the basic concepts and background in both computational intelligence and EDA fields. Their relationships are discussed and the challenging problems which will be addressed in this book are introduced.

The main content of this book, Chaps. 2–10, can be divided into three parts.

The first part includes Chaps. 2–4, focusing on the global optimization of highly constrained problems.

Chapter 2 introduces the basics or fundamentals of evolutionary algorithms and constraint handling methods with the practical application of analog integrated circuit sizing. This chapter covers evolutionary algorithms for single and multi-objective optimization and basic constraint handling techniques. Popular methods are introduced with practical examples.

Chapter 3 discusses advanced techniques for high performance design optimization. This chapter reviews advanced constraint handling methods and hybrid methods and introduces some popular methods. Practical examples are also provided.

Chapter 4 introduces optimization problems with fuzzy constraints to integrate the humans' flexibility and high optimization ability of evolutionary algorithms. Fuzzy sets, fuzzy constraint handling methods and the integration of fuzzy constraint handling methods into previous techniques are presented. The application field is fuzzy analog circuit sizing.

The second part includes Chaps. 5 and 6, and focuses on efficient global optimization in uncertain environments, or robust design optimization.

Chapter 5 provides an overview of uncertain optimization, and the application area: variation-aware analog circuit sizing. Two common efficiency enhancement methods for uncertain optimization are then introduced, including some basics of computational statistics.

Chapter 6 introduces ordinal optimization-based efficient robust design optimization methods. The method to cooperate ordinal optimization with hybrid methods, single and multi-objective constrained optimization methods is then discussed with practical examples.

The third part includes Chaps. 7–10, and focuses on efficient global optimization of computationally expensive black-box problems.

Chapter 7 reviews surrogate model assisted evolutionary algorithms and the application area: design automation of mm-wave integrated circuits and complex antennas. Two machine learning methods, Gaussian process and artificial neural networks are introduced.

Chapter 8 introduces the fundamentals of surrogate model assisted evolutionary algorithms that are applied to high-frequency integrated passive component synthesis. Three popular methods to handle the prediction uncertainty, which is the fundamental problem when integrating machine learning techniques with evolutionary algorithms, are introduced with practical examples.

Chapter 9 introduces a method for mm-wave linear amplifier design automation. The methods to analyze the problem from the computation aspect, to utilize its properties and to transform it to a problem that can be solved by the techniques introduced in Chap. 8 are discussed. Instead of introducing new computational intelligence techniques, this chapter concentrates on how to make use of the basic techniques to solve complex problems.

Chapter 10 focuses on the cutting-edge problem in surrogate model assisted evolutionary algorithms: handling of high dimensionality. Two state-of-the-art techniques, dimension reduction and surrogate model-aware evolutionary search mechanism are introduced. The practical examples are the synthesis of mm-wave nonlinear integrated circuits and complex antennas.

Finally, we would like to thank the Alexander von Humboldt Foundation, Professor Guenter Rudolph, Professor Helmut Graeb, Professor Tom Dhaene, Professor Qingfu Zhang, Professor Guy A. E. Vandenbosch, Dr. Trent McConaghy, Dr. Patrick Reynaert, Dixian Zhao, Dr. Hadi Aliakbarian, Dr. Brecht Machiels, Zhongkun Ma, Noel Deferm, Wan-ting Lo, Bohan Yang, Borong Su, Chao Li, Jarir Messaoudi, Xuezhi Zheng and Ying He. We also express our appreciation to Professor Janusz Kacprzyk and Dr. Thomas Ditzinger for including this book in the Springer series on "Studies in Computational Intelligence".

Bo Liu
Francisco V. Fernández
Georges Gielen

Contents

1   **Basic Concepts and Background**   1
     1.1   Introduction   1
     1.2   An Introduction into Computational Intelligence   5
         1.2.1   Evolutionary Computation   5
         1.2.2   Fuzzy Logic   7
         1.2.3   Machine Learning   9
     1.3   Fundamental Concepts in Optimization   9
     1.4   Design and Computer-Aided Design of Analog/RF IC   11
         1.4.1   Overview of Analog/RF Circuit
               and System Design   11
         1.4.2   Overview of the Computer-Aided Design
               of Analog/RF ICs   13
     1.5   Summary   15
     References   16

2   **Fundamentals of Optimization Techniques
in Analog IC Sizing**   19
     2.1   Analog IC Sizing: Introduction and Problem Definition   19
     2.2   Review of Analog IC Sizing Approaches   21
     2.3   Implementation of Evolutionary Algorithms   23
         2.3.1   Overview of the Implementation of an EA   23
         2.3.2   Differential Evolution   24
     2.4   Basics of Constraint Handling Techniques   27
         2.4.1   Static Penalty Functions   27
         2.4.2   Selection-Based Constraint Handling Method   28
     2.5   Multi-objective Analog Circuit Sizing   29
         2.5.1   NSGA-II   29
         2.5.2   MOEA/D   32
     2.6   Analog Circuit Sizing Examples   34
         2.6.1   Folded-Cascode Amplifier   34
         2.6.2   Single-Objective Constrained Optimization   34
         2.6.3   Multi-objective Optimization   36
     2.7   Summary   38
     References   39

3   High-Performance Analog IC Sizing: Advanced Constraint
**   Handling and Search Methods** 41
     3.1   Challenges in Analog Circuit Sizing 41
     3.2   Advanced Constrained Optimization Techniques 42
           3.2.1   Overview of the Advanced Constraint
                   Handling Techniques 42
           3.2.2   A Self-Adaptive Penalty Function-Based Method ... 44
     3.3   Hybrid Methods 47
           3.3.1   Overview of Hybrid Methods 47
           3.3.2   Popular Hybridization and Memetic Algorithm
                   for Numerical Optimization 48
     3.4   MSOEA: A Hybrid Method for Analog IC Sizing 50
           3.4.1   Evolutionary Operators 50
           3.4.2   Constraint Handling Method 53
           3.4.3   Scaling Up of MSOEA 53
           3.4.4   Experimental Results of MSOEA 56
     3.5   Summary 61
     References 61

4   Analog Circuit Sizing with Fuzzy Specifications:
**   Addressing Soft Constraints** 63
     4.1   Introduction 63
     4.2   The Motivation of Analog Circuit Sizing
           with Imprecise Specifications 64
           4.2.1   Why Imprecise Specifications Are Necessary 64
           4.2.2   Review of Early Works 65
     4.3   Design of Fuzzy Numbers 66
     4.4   Fuzzy Selection-Based Constraint Handling
           Methods (Single-Objective) 68
     4.5   Single-Objective Fuzzy Analog IC Sizing 70
           4.5.1   Fuzzy Selection-Based Differential
                   Evolution Algorithm 70
           4.5.2   Experimental Results and Comparisons 71
     4.6   Multi-objective Fuzzy Analog Sizing 75
           4.6.1   Multi-objective Fuzzy Selection Rules 76
           4.6.2   Experimental Results for Multi-objective
                   Fuzzy Analog Circuit Sizing 78
     4.7   Summary 81
     References 82

5   Process Variation-Aware Analog Circuit Sizing:
**   Uncertain Optimization** 85
     5.1   Introduction to Analog Circuit Sizing Considering
           Process Variations 85

5.1.1 Why Process Variations Need to be Taken
into Account in Analog Circuit Sizing 85
5.1.2 Yield Optimization, Yield Estimation
and Variation-Aware Sizing.................... 86
5.1.3 Traditional Methods for Yield Optimization 88
5.2 Uncertain Optimization Methodologies 90
5.3 The Pruning Method........................... 92
5.4 Advanced MC Sampling Methods 93
5.4.1 AYLeSS: A Fast Yield Estimation Method
for Analog IC 95
5.4.2 Experimental Results of AYLeSS............... 99
5.5 Summary.................................... 103
References 103

6 Ordinal Optimization-Based Methods for Efficient
Variation-Aware Analog IC Sizing 107
6.1 Ordinal Optimization 108
6.2 Efficient Evolutionary Search Techniques 110
6.2.1 Using Memetic Algorithms 110
6.2.2 Using Modified Evolutionary Search Operators 111
6.3 Integrating OO and Efficient Evolutionary Search 113
6.4 Experimental Methods and Verifications of ORDE......... 116
6.4.1 Experimental Methods for Uncertain Optimization
with MC Simulations 116
6.4.2 Experimental Verifications of ORDE 117
6.5 From Yield Optimization to Single-Objective Analog
Circuit Variation-Aware Sizing 119
6.5.1 ORDE-Based Single-Objective Variation-Aware
Analog Circuit Sizing 120
6.5.2 Example 121
6.6 Bi-objective Variation-Aware Analog Circuit Sizing........ 122
6.6.1 The MOOLP Algorithm 123
6.6.2 Experimental Results 128
6.7 Summary.................................... 130
References 130

7 Electromagnetic Design Automation: Surrogate Model
Assisted Evolutionary Algorithm....................... 133
7.1 Introduction to Simulation-Based Electromagnetic
Design Automation............................. 134
7.2 Review of the Traditional Methods................... 135
7.2.1 Integrated Passive Component Synthesis 135
7.2.2 RF Integrated Circuit Synthesis 137
7.2.3 Antenna Synthesis 138

7.3   Challenges of Electromagnetic Design Automation.   139
7.4   Surrogate Model Assisted Evolutionary Algorithms   140
7.5   Gaussian Process Machine Learning   142
       7.5.1   Gaussian Process Modeling   143
       7.5.2   Discussions of GP Modeling   144
7.6   Artificial Neural Networks.   147
7.7   Summary.   148
References   149

8   Passive Components Synthesis at High Frequencies:
Handling Prediction Uncertainty   153
8.1   Individual Threshold Control Method   154
       8.1.1   Motivations and Algorithm Structure   154
       8.1.2   Determination of the MSE Thresholds   155
8.2   The GPDECO Algorithm.   158
       8.2.1   Scaling Up of GPDECO   158
       8.2.2   Experimental Verification of GPDECO.   160
8.3   Prescreening Methods   161
       8.3.1   The Motivation of Prescreening   161
       8.3.2   Widely Used Prescreening Methods   163
8.4   MMLDE: A Hybrid Prescreening and Prediction Method   165
       8.4.1   General Overview.   165
       8.4.2   Integrating Surrogate Models into EA.   166
       8.4.3   The General Framework of MMLDE   168
       8.4.4   Experimental Results of MMLDE   169
8.5   SAEA for Multi-objective Expensive Optimization
       and Generation Control Method   173
       8.5.1   Overview of Multi-objective Expensive
                Optimization Methods.   174
       8.5.2   The Generation Control Method.   175
8.6   Handling Multiple Objectives in SAEA.   176
       8.6.1   The GPMOOG Method.   177
       8.6.2   Experimental Result   180
8.7   Summary.   182
References   182

9   mm-Wave Linear Amplifier Design Automation:
A First Step to Complex Problems   185
9.1   Problem Analysis and Key Ideas   186
       9.1.1   Overview of EMLDE   186
       9.1.2   The Active Components Library and the Look-up
                Table for Transmission Lines.   187
       9.1.3   Handling Cascaded Amplifiers.   188
       9.1.4   The Two Optimization Loops   188

9.2   Naive Bayes Classification 190
9.3   Key Algorithms in EMLDE......................... 191
        9.3.1   The ABGPDE Algorithm.................... 191
        9.3.2   The Embedded SBDE Algorithm 193
9.4   Scaling Up of the EMLDE Algorithm................. 193
9.5   Experimental Results 195
        9.5.1   Example Circuit......................... 195
        9.5.2   Three-Stage Linear Amplifier Synthesis 197
9.6   Summary.................................... 199
References 199

10   mm-Wave Nonlinear IC and Complex Antenna Synthesis:
     Handling High Dimensionality 201
     10.1   Main Challenges for the Targeted Problem
             and Discussions 202
     10.2   Dimension Reduction 204
             10.2.1   Key Ideas 204
             10.2.2   GP Modeling with Dimension Reduction
                       Versus Direct GP Modeling................. 206
     10.3   The Surrogate Model-Aware Search Mechanism 206
     10.4   Experimental Tests on Mathematical Benchmark Problems ... 210
             10.4.1   Test Problems 210
             10.4.2   Performance and Analysis.................. 210
     10.5   60 GHz Power Amplifier Synthesis by GPEME.......... 219
     10.6   Complex Antenna Synthesis with GPEME.............. 223
             10.6.1   Example 1: Microstrip-fed Crooked
                       Cross Slot Antenna...................... 225
             10.6.2   Example 2: Inter-chip Wireless Antenna 228
             10.6.3   Example 3: Four-element Linear Array Antenna 230
     10.7   Summary.................................... 232
     References 234

Chapter 1
Basic Concepts and Background

1.1 Introduction

Computational intelligence (CI) is a branch of artificial intelligence (AI). The goal of AI is to understand how we think and then go further to build intelligent entities [1]. For instance, scientific research is carried out by scientists at present. AI, however, aims at building a machine system that can do research like several expert researchers. An intelligent agent is expected to have the following abilities: thinking and reasoning, processing knowledge, learning, planning and scheduling, creativity, motion and manipulation, perception and communication. With those capabilities, it could act as a machine brain. CI is the part of the machine brain focusing on reasoning, learning and planning. They are defined as nature-inspired methodologies to solve complex computational problems for which traditional mathematical methodologies are ineffective, e.g., the optimization of a non-differentiable, non-convex function, the automatic control of a bicycle, the prediction of new outputs to untested inputs only based on a given experimental data set, etc. CI mainly includes Evolutionary Computation (EC) for global optimization, which mimics the biological evolution, Artificial Neural Networks (ANNs) for machine learning, which mimics the signal processing in human brain and Fuzzy Logic for reasoning under uncertainty, which mimics the reasoning of the human being. First, CI techniques were developed from 1940 to 1970, and have been widely applied to real-world problems since 1990. Since the 1950s, the increasing complexity of industrial products has created a rapidly growing demand for automated problem solving. The growth rate of the research and development capacity could not keep pace with these needs. Hence, the time available for thorough problem analysis and tailored algorithm design has been and is still decreasing. This trend implies an urgent need for robust and general algorithms with satisfactory performance [2]. Undoubtedly, CI provides an answer to the above challenge. Nowadays, CI techniques play an important role in many industrial areas, from chemical engineering to bioinformatics, from automobile design to intelligent transport systems, from aerospace engineering to nano-engineering.

   This book introduces CI techniques for electronic design automation (EDA). In the semiconductor industry, the pace of innovation is very high. Over the past four

B. Liu et al., *Automated Design of Analog and High-frequency Circuits*,
Studies in Computational Intelligence 501, DOI: 10.1007/978-3-642-39162-0_1,
© Springer-Verlag Berlin Heidelberg 2014

decades, the number of transistors on a chip has increased exponentially in accordance with Moore's law [3, 4]. Moreover, the development speed is even higher in recent years: integrated circuits (IC), serve as the foundation of the information age, improving our life in a number of ways: fast computers, cell phones, digital televisions, cameras. In this decade, more social needs, such as health, security, energy, transportation, will benefit from the "more than Moore" development. Instead of digital ICs, this book concentrates on the design automation methodologies of analog ICs, high-frequency ICs and antennas. Besides a general review, special attention is paid to the new challenging problems appeared in recent years. To address these challenges, state-of-the-art novel algorithms based on CI techniques are introduced, some of which are also cutting-edge research topics in the CI field. Let us first see the challenges we are facing from the EDA point of view.

- Challenge on high-performance analog IC design
  Driven by the market demands and advances in IC fabrication technologies, the specifications of modern analog circuits are becoming increasingly stringent. On the other hand, with the scaling down of device sizes, the transistor equivalent circuit models for manual design often yield low accuracy, while the SPICE models, which are very accurate, are too complex to be used for manual designers. Hence, the design of high-performance analog ICs in a limited amount of time is not an easy task even for skilled designers. When we face an analog cell with 20–50 transistors, the performance optimization is more difficult. Modern numerical optimization techniques have been introduced to analog IC sizing, but the objective optimization and constraint handling abilities of most of the existing methods are still not good enough for high-performance analog IC sizing [5].
- Process variations become a headache with the scaling down of the transistors
  Industrial analog integrated circuit design not only calls for fully optimized nominal design solutions, but also requires high robustness and yield in the light of varying supply voltage and temperature conditions, as well as inter-die and intra-die process variations [6]. With the scaling down of the transistors, the variation becomes larger and larger and will continue to get worse in the future technologies. As stated in the ITRS reports [4], the variation of the threshold voltage of a transistor reached 40 % in 2011 and is predicted to reach 100 % in the coming 10 years. Nowadays, even a single atom out of place may worsen the circuit behavior or even make the circuit fail. Figure 1.1 shows variability-induced failure rates for three simple canonical circuit types with the shrinking of the technology. Therefore, high-yield design is highly needed with shrinking device sizes. Designer often introduces some over-design to take the degradation of performances brought by process variations into account. However, this may lower the performances or require more power and / or area. In recent years, some variation-aware analog IC sizing methods based on computational intelligence techniques have been proposed, such as [7]. But many of the existing methods have the problems of not being general enough, not accurate enough, not applicable to modern technologies or not fast enough [7, 8].
- High-data-rate communication brings challenges to mm-wave circuit design

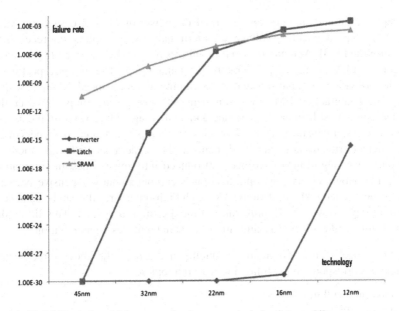

Fig. 1.1 Variability-induced failure rates for three canonical circuit types [4]

Radio frequency (RF) integrated circuits are adding a dramatic impact to our life. Today we can access voice / data and entertainment in virtually every corner of the globe, from short-range Bluetooth communication to satellite networks. Hence, high-data-rate communication channels are needed. According to Shannon's theorem, one effective way to increase the data rate is to use more bandwidth. The information-bearing signal is usually modulated around a carrier frequency for proper propagation. Therefore, more bandwidth is available around higher carrier frequencies [9]. This leads to the dramatic increase of research and applications of mm-wave RF ICs.

At mm-wave frequencies, the passive components equivalent circuit models which are often used by manual designers for low-GHz RF ICs are not accurate enough anymore due to the distributed effects. Because of this, maybe the only way left for the designers is to directly rely on the S-parameters from electromagnetic (EM) simulations and design the circuit based on "experience and trial". This makes designing high-performance mm-wave RF ICs more difficult and requires highly-skilled designers. Even for skilled designers, design for high performance is not an easy task. In CAD tools for RF ICs, most existing methods only support low-GHz RF IC synthesis [10–12].

In communications, antennas are equally important as integrated circuits: they convert electric currents into radio waves, and viceversa. Antennas are used both in the transmitter and receiver. In transmission, a radio transmitter applies an oscillating radio frequency electric current to the antenna's terminals, and the antenna radiates the energy from the current as electromagnetic waves (radio waves). In reception, an antenna intercepts some of the power of an electromagnetic wave in order to produce a tiny voltage at its terminals, that is applied to a receiver to be amplified [13]. Antenna research mainly focuses on two aspects: (1) Antenna theory and EM simulation, (2) Practical antenna design. The former aspect has a high correlation with physics and is developing at a high speed. Many new discoveries and practical EM simulation methodologies are emerging. However, the latter aspect is still in an "experience and trial" stage. Only very simple antennas have systematic design methodologies. For complex antennas, designers often first detect critical parameters and then use the "experience and trial" method. Evolutionary algorithms have been introduced into antenna design automation [14, 15]. However, antenna optimization is computationally expensive because of the embedded EM simulations. Although high-quality results can be achieved, several days to weeks of optimization time are often necessary. This limits the applicability of evolutionary algorithm-based antenna design automation.

With respect to the computational intelligence aspect, the above three design challenges correspond to the following research topics:

- Constraint handling
- Hybrid methods
- Uncertainty optimization
- Computationally expensive black-box optimization

High-performance analog IC design can be addressed with advanced evolutionary computation algorithms research. Most evolutionary algorithms (EAs) are constraint blind, but high-performance analog IC sizing requires that the optimization handles severe constraints. Also, hybrid methods are often necessary to obtain highly optimized solutions. These two issues have already received much attention in the computational intelligence field [16, 17]. The goal here is to improve the "effectiveness". However, improving the "effectiveness" is not the main goal of variation-aware analog IC sizing, mm-wave IC and antenna synthesis. Instead, "efficiency" is the key objective. Variation-aware analog IC sizing focuses on the efficient solution of uncertain optimization; mm-wave IC and antenna synthesis focuses on the efficient solution of small-and medium-scale computationally expensive black-box optimization. Note that this does not mean that we do not want highly optimized designs, but efficiency improvement to these problems is more critical to be able to complete the design in an acceptable time (from hours up to several days). For these research areas, although computational intelligence researchers have made their promising progress, they are still in their infancy stage and many fundamental issues need to be addressed.

Clearly, solving these problems is crucial to both the electronic design automation field and to the computational intelligence field. Hence, this book is both for EDA and CI researchers / engineers. For the former readers, the book presents: advanced methods for sizing analog ICs with severe performance requirements, efficient and general methods for analog IC yield optimization, multi-objective variation-aware analog IC sizing and general methods for synthesis of integrated passive components, linear and nonlinear RF amplifiers at mm-wave frequencies and complex antennas in a very practical time. For the latter readers, this book introduces CI methods of constraint handling, hybrid methods, uncertain optimization for robust design and surrogate model assisted evolutionary algorithms. Both tutorials for beginners and state-of-the-art methods are included.

This book is organized in such a way that its contents provide a general overview of the focused research issues from both fields. This chapter provides a short introduction to CI techniques and the concepts used in optimization research, together with introductions to the background of the application area, IC design and computer-aided design (CAD). From Chap. 2, this book is divided into three parts. Chaps. 2–4 cover **high-performance analog IC sizing**. The CI techniques introduced in this part are evolutionary algorithms for single and multi-objective optimization, constraint handling techniques, hybrid methods and fuzzy sets. Practical CAD methods for analog IC sizing with severe specifications, fuzzy design specifications are presented. The goal is the effectiveness enhancement of the evolutionary algorithms. Chapters 5–7 cover **variation-aware analog IC sizing**. The CI techniques introduced in this part are optimization in uncertain environments. In addition, an introduction to fast Monte-Carlo simulation, which is often used in uncertain optimization, is provided. Practical CAD methods for fast yield estimation and fast and accurate single- and multi-objective variation-aware analog IC sizing are presented. Chapters 8–11 cover **synthesis using computationally expensive EM simulations**. The CI techniques introduced in this part are machine learning methods and surrogate model-assisted evolutionary algorithms. The machine learning methods introduced in this book is not limited to ANN, and Gaussian Process (GP) machine learning, which is a state-of-the-art method, is paid special attention. Practical CAD methods for fast and high-quality synthesis for integrated passive components, linear amplifiers, non-linear amplifiers and complex antennas are presented. The goal of Chaps. 5–11 is the efficiency enhancement of the evolutionary algorithms.

The remainder of this chapter is organized as follows. Section 1.2 provides a brief introduction to evolutionary computation, fuzzy sets and machine learning. Because optimization is the focus throughout this book, some basic concepts of optimization research are illustrated in Sect. 1.3. Section 1.4 gives the background into analog and RF IC design and CAD. Section 1.5 summarizes this chapter.

1.2 An Introduction into Computational Intelligence

1.2.1 Evolutionary Computation

Evolutionary computation algorithms mimic the biological mechanisms of evolution to approximate global optimal point of a problem. In the 1950s, the use of Darwinian principles, "survival of the fittest", to automated problem solving has appeared, which is the origin of evolutionary computation (EC). It is not surprising that some computer scientists choose natural evolution as a source of inspiration, as it is evidenced to be successful in a variety of aspects. EC is composed of genetic algorithms (GAs), evolution strategies (ES) and evolutionary programmings (EPs), collectively known as evolutionary algorithms (EAs) [18]. The common elements of EC algorithms are the use of a population of individuals, which are processed by a set of operators and are evaluated by the fitness function(s). The operators consist of reproduction, random variation and selection. The purpose of these operators is to generate candidates with higher and higher fitness value. The selection is based on the fitness function, which represents "how well" the candidate is in achieving the goals and which decides the probability of survival. This is a "generate and test" problem solving process, and the candidates with higher fitness values have more chance to be kept and used as parents for constructing further candidate solutions. Figure 1.2 shows the basic concepts in EA [19].

The general process of an EA algorithm is shown in Fig. 1.3. The population is first randomly initialized. Then, crossover takes place by cutting and recombining pieces of binary strings of different parent candidates (considering a binary GA [18]) to generate new child candidates. The purpose of this step is to re-construct or recombine the information from the parents. If beneficial information is recombined, the fitness value of the new candidate has a higher probability to be good. There are many kinds of crossover operators, such as single-point or multi-point crossover. The next step is mutation, whose purpose is to introduce new information to the population to enable global exploration. For example, for bitwise mutation, a "0" may be changed to a "1" for a binary string. After mutation, the child candidates are generated. Selection effectively gives a higher probability to candidates with higher fitness values to contribute to the succeeding generations. For example, for the commonly used roulette-wheel selection, probabilities are assigned to different candidates according to their fitness values. Tournament selection, on the other hand, randomly picks two candidates and selects the one with higher fitness value to enter

Fig. 1.2 Key concepts in evolutionary algorithms (from [19])

Algorithm $\left\{\begin{array}{l}\end{array}\right.$
- ➤ Population: Set of Individuals (Solutions)
- ➤ Parent: Member of Current Generation
- ➤ Children:Members of Next Generation
- ➤ Generation: Successively Created Populations
    (EA Iteration)

Data Structure $\left\{\begin{array}{l}\end{array}\right.$
- ➤ Chromosome: Solution's Coded Form; Vector
    (String) Consists of Genes With Alleles
    Assigned
- ➤ Fitness: Number Assigned to a Solution;
    Represent's "Desirability"

Fig. 1.3 General flow of
standard GA

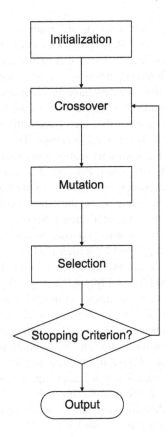

the next generation until all the places are filled. If the stopping criterion is met, such as a maximum number of generations, the best solution so far will be the output; otherwise, a new iteration begins.

Besides GA, there are many other EAs. Some popular ones are differential evolution (DE) [20], particle swarm optimization [21] and ant colony algorithm [22], which are proven to be very effective in real-world applications. They are different from GA in many aspects, but still use the evolution framework.

Compared with traditional methods such as the Newton method, many real-world problems, such as those that are non-convex, discontinuous and non-differentiable, fall into the playfield of EC. Another advantage of EA is that they have the ability of global search, aiming at finding the globally optimal point of a function.

1.2.2 Fuzzy Logic

The real world is complex and full of uncertainty, which is unavoidable in most practical engineering applications. However, the computer itself does not have the ability to handle uncertain information. For example, when we ask a question "what is the criterion to define a tall man?", one person may think that a tall person should be higher than 5 feet, while another people may think that a tall person should be higher than 6 feet. The concept of "tall" is different from person to person. However, although they have different understanding of "a tall man", the two persons can also understand and communicate to each other on the question of if a man is tall. The reason is that the concept of "tall" has similar meanings from the point of view of two different persons. The computer, in contrast, has difficulties in doing this. The computer requires an exact threshold to define "tall", and then judges if a man is tall or not. For complex problems, the human brain has the ability of imprecise reasoning even when it only has very limited information. To make the computer more intelligent, the same modeling method as the human brain has been introduced into computer science for modeling complex systems. That is fuzzy logic [23].

The most important concept in fuzzy logic is fuzzy sets. The idea of memberships suggested by Zadeh [24] is the key to decision making when facing non-random uncertainty. Here, we again take the example of height to illustrate fuzzy sets, as shown in Fig. 1.4. Suppose that 6 feet is the precise definition of "tall". A crisp set has a unique membership function, such as defining that 5 to 7 feet is tall, and that a 4.999 feet man is not tall. On the other hand, a fuzzy set defines a region around 6 feet as tall, which is imprecise, or fuzzy. In this region, each height has a membership value between 0 and 1 (infinite membership functions), which describes the degree it conforms to the concept of "tall". 6 feet has full membership 1, and 3 feet has 0 membership, which is the same as for the crisp set. Between 4.8 and 7.2, the membership is between 0 and 1, the closer to 6 feet, the more close to the definition of "tall". For 7 feet, it has a low membership value of "tall", but maybe a higher membership value of "very tall". In fuzzy reasoning, one point often has several membership values defined by different sets and is judged by all of them.

We use fuzzy sets to perform fuzzy reasoning. The rules of reasoning are often linguistic rules, such as: if a man is tall, and if a man can move swiftly, then he has good potential to be a basketball player. All the concepts used here are defined by fuzzy sets, and through fuzzy sets calculations, the resulting membership can be obtained and the corresponding rules are selected. At last, a decision can be made by defuzzification.

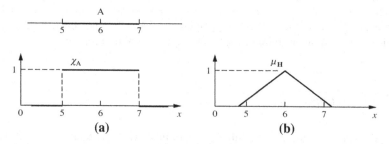

Fig. 1.4 Height membership function for **a** crisp set, **b** fuzzy set (from [23])

1.2.3 Machine Learning

Fuzzy reasoning is based on a set of decision rules, which are predefined. An intelligent agent not only can act based on the rule set, but can also automatically develop the rule set, which is the process of learning. Observation from the outside world decides "what to learn" and its own decision-making process decides "how to learn". A machine learning mechanism often contains a performance element telling the machine what action to take and a learning element modifying the previous element to improve the decisions. Machine learning can mainly be classified into supervised learning, unsupervised learning and reinforcement learning [1]. This book only covers supervised learning for classification and prediction.

Classification and prediction play significant roles in data mining and have been widely used in many science and engineering areas. Classification is the problem of identifying the classification to which new observations belong to, when the identity of the classifications of the new observations is unknown, on the basis of a training data set containing observations with known classifications. Prediction is the problem of predicting the function values of a set of not evaluated inputs, on the basis of a training data set of already evaluated inputs and their function evaluations. In bioinformatics, the DNA sequences are classified by intelligent machines. In communication, classification plays an important role in speech and handwriting recognition. In economy, there is much stock market analysis software based on predictions. In this book, statistical learning-based prediction techniques are used to learn functions which consume a large amount of computational time. In mm-wave IC design automation, evaluating one design candidate requires a set of computationally expensive EM simulations. When directly using canonical EAs, the synthesis would require an intractable time. Therefore, statistical machine learning methods are used to learn and predict the function value on-line (active learning) to enhance the optimization efficiency. That is the research issue of surrogate model assisted evolutionary algorithm (SAEA).

1.3 Fundamental Concepts in Optimization

Electronic design automation problems are typically cast as optimization problems. Fundamental concepts of optimization used in the following chapters are introduced in this Section.

- Single-objective optimization
  An unconstrained single-objective optimization problem can be described by (1.1):

$$\begin{aligned}
&\text{minimize } f(x) \\
&\text{s.t.} \qquad x \in [a, b]^d.
\end{aligned} \tag{1.1}$$

where x is the vector of decision variables; d is the dimension of x; $[a, b]^d$ are the search ranges of the decision variable x, $f(x)$ is the objective function and "s.t." is the abbreviation of "subject to". The optimal solution is the value of $x' \in [a, b]^d$ where there is no other point $x^* \in [a, b]^d$ with $f(x^*) < f(x')$ (considering a minimization problem).

- Multi-objective optimization
An unconstrained multi-objective optimization problem can be described by (1.2):

$$\text{minimize } \{f_1(x), f_2(x), \ldots, f_m(x)\}$$
$$\text{s.t.} \qquad x \in [a, b]^d. \tag{1.2}$$

where x is the decision variable, d is the dimension of x, $[a, b]^d$ are the search ranges of the decision variable x. $f_1(x), f_2(x), \ldots, f_m(x)$ are the optimization objectives. According to different purposes and requirements in the decision-making process, multi-objective optimization techniques can roughly be classified into two categories [19]: (1) priori methods: a decision maker specifies his preferences on these objectives and thus transforms the multi-objective problem into a single-objective one by using aggregation methods (e.g., summation with weighting coefficients), and (2) posteriori methods: they produce a number of well representative optimal trade-off candidate solutions for a decision maker to check. A Pareto optimal solution is a candidate solution that achieves the best trade-off. There can be many, even infinitely many, Pareto optimal solutions to a multi-objective optimization problem (MOP). The set of all the Pareto optimal solutions is called the Pareto set (PS) and its image in the objective space is the Pareto front (PF).

- Constrained optimization
A constrained optimization problem can be described by:

$$\text{minimize } f(x)$$
$$\text{s.t.} \qquad g_i(x) \le 0, i = 1, 2, \ldots, k. \tag{1.3}$$
$$x \in [a, b]^d.$$

where $f(x)$ is the optimization goal, and $g_i(x)$ are the constraints. The difference between the optimization goal and the constraints is as follows. Considering a single-objective minimization problem, for $f(x)$ the smaller the function value, the better the candidate solution is. On the other hand, for a constraint $g_1(x) \le 0$, a candidate solution with $g_1(x) = -10$ is not better than that with a $g_1(x) = -5$, but $g_1(x) = 1$ violates the constraint, and is an infeasible solution. Traditionally, the goal of a constrained optimization problem is to find the best solution(s) of the objective function(s) under the condition of all constraints being satisfied. (1.3) shows a single-objective constrained optimization problem but constrained optimization problems can also have multiple objectives. For multi-objective constrained optimization, when satisfying all the constraints, approximating the PF is the goal of the optimization.

- Stochastic / uncertain optimization

Stochastic optimization can have a single or multiple objectives, with or without constraints. The point to identify this kind of optimization problems is that at least an objective or a constraint involves random or fuzzy variables. Therefore, different evaluations of that objective or constraint with the same decision variables receive different results. There are several methods to handle uncertain objectives or constraints, such as expected value, chance value, dependent chance value [25], which will be addressed in the following chapters.

- Computationally expensive optimization
  Computationally expensive optimization can be in any category of optimization problems defined above. The main characteristic is that the function evaluation (either objectives or constraints) is very time consuming. The high computational cost can be caused by many reasons. For example, for analog circuit yield optimization, the calculation of yield often needs a large number of statistical samples. Although the computational effort spent on each sample is not long, the accumulated time of all the samples is long. For EM simulation, each candidate only needs one simulation, but the EM simulation itself costs a long time. In the following chapters, single- and multi-objective expensive optimization problems with or without constraints will be introduced.
- Benchmark test problems
  Benchmark test problems [26] are very important in intelligent optimization research. In the EC field, it is common to test and compare different algorithms using a set of benchmark functions. Benchmark functions are mathematical functions with different characteristics. Many of them are complex due to non-differentiability, multi-modality (more than one local optima), high dimensionality and non-separability (cannot be re-written as a sum of a set of functions with just one variable). Although they are difficult to optimize, their globally optimal solution(s) is / are often known, so as to evaluate an algorithm quantitatively. Benchmark problems will be mentioned and used in this book.

1.4 Design and Computer-Aided Design of Analog/RF IC

1.4.1 Overview of Analog/RF Circuit and System Design

We will now introduce how a chip is made from zero. A large analog / RF system with many transistors and passive components is not designed as a whole, but is decomposed into sub-blocks. Each sub-block will be further decomposed down to the cell level. An analog / RF cell is a small circuit having a certain basic function, such as an amplifier, a mixer. The decomposition methods can be bottom-up or top-down [27]. Top-down decomposition firstly designs the top level, which results in the specifications of the sub-blocks. Then, the sub-blocks are designed, which at the same time results in the specifications of the lower-level sub-blocks. The process continues until the lowest level is designed. Bottom-up decomposition, on the contrary, firstly

Fig. 1.5 Analog cell design flow

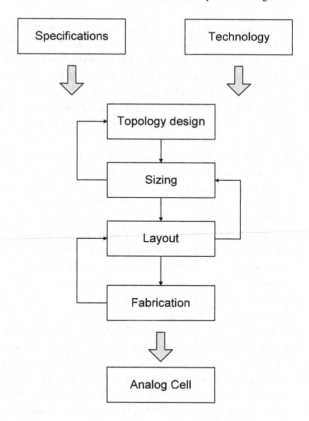

generates Pareto-optimal surfaces from the lowest level blocks, which provides the design space of the higher level. This process continues until the top level is designed. The bottom-up design needs multi-objective optimization, which can only be done by CAD tools [7]. This book does not focus on CAD tools for hierarchical system design, but focuses on CAD methods for the design of analog / RF cells.

Figure 1.5 shows the design flow of an analog cell. The inputs of the system are the design specifications. According to these specifications and the technology used, the designer firstly decides on the topology, or circuit configuration. He / She may choose an existing one, modify an existing one, or propose a new topology. After the topology is decided, the parameters will be designed, including transistor sizes, capacitor values, biasing voltages or currents. The goal of this step is to make the parameterized circuit topology satisfy the specifications as verified by circuit simulation. If the simulation results do not meet the design specifications, re-sizing is necessary or even the circuit topology needs to be changed. If the simulation results meet the specifications, the layout will be created. Layout is to express the parametric design into physical design, whose goal is to create the shapes for fabrication. Parasitics are introduced in the layout process because the non-ideal effects of silicon materials and interconnects have not been considered in sizing. Hence, the post layout verification is necessary. The parasitics are extracted and a more complex simulation including the parasitics is then performed. Again, if the specifications are not met, a redesign cycle is needed. If the post-layout simulation result meets the specifications, we can fabricate it into a real chip. However, it is not guaranteed that the fabricated chip can work. If the measurement result does not meet the specifications, we need to find the problem and re-design. The whole process is often time consuming.

The RF IC design flow has similarities with the analog IC design flow described in Fig. 1.5. One of the main differences is that a large amount of effort needs to be spent on the integrated passive components (e.g., inductor, transformer, instead of transistors), especially at mm-wave frequencies. At such frequencies, the wavelengths of signals are roughly the same as the dimensions of the devices and the lumped-element circuit theory is inaccurate. Due to the distributed effects, most of the equivalent circuit models are narrow-banded. This means that one can build a good parasitic-aware equivalent circuit model at 60 GHz, but the same model cannot describe the behavior of the passive component at 70 GHz with reasonable accuracy [28]. For low-frequency analog ICs, equivalent circuit models for the devices are crucial in manual design, but in RF IC design, they are often not available. Hence, the designers are forced to rely on experience, intuition and inefficient simulators to predict the circuit performance and update the design parameters.

1.4.2 Overview of the Computer-Aided Design of Analog/RF ICs

Over the years, as the technology becomes more and more complex, it is not possible to design a circuit without CAD tools. CAD tools can be classified into five main classes: analysis tools, insight tools, editors, verification tools and design automation tools. Among them, analysis tools and design automation tools are the most important. Analysis tools can be regarded as the "evaluator", which provides the performance of a given design. They work in a "forward" way. The design automation tools work in a "backward" way, and they provide the (best) design for a given set of specifications.

- Analysis tools
  The most famous analysis tool for analog ICs is the SPICE circuit simulator [29, 30], which has proven to be essential in both manual design and CAD. Almost no IC design or CAD work does not use SPICE for circuit simulation (evaluation). Moreover, the dramatically increased, and still increasing, computing power over the years has enabled the "simulation in the loop" design automation methodologies. The inputs of SPICE include the technology device model parameters, the netlist and the requested analysis types. The technology model replaces the devices by accurate models. The netlist contains all the information of the circuit design, such as the topology, device sizes and biasing voltages or currents. Kirchoff's law is used to analyze the circuit, obtaining the voltages at all nodes and the currents in all the devices. The performance results are then calculated. DC, AC and transient

analysis are commonly used for analog ICs. DC analysis provides the biasing conditions of all the devices. AC analysis presents small-signal performances over frequencies. Transient analysis reports the behavior over time. Using these analyses, most of the performances of an analog IC can be evaluated.

There are also useful tools to help post-layout simulation. Mentor Graphics Resistor-Capacitor Extraction (RCX) tool in Calibre [31] is an example. This tool generates a new netlist with all the layout-induced parasitics to be included in the SPICE simulation.

The main analysis tools for RF IC design are electromagnetic (EM) analysis tools, small signal analysis tools and steady-state analysis tools. At high frequencies, especially mm-wave frequencies, even the parasitic-aware equivalent circuit models for passive devices are no longer accurate. Therefore, EM simulation is needed for the passive components. The function of the EM simulation tools is to calculate S-parameters for a passive component. Many electrical properties of networks of passive components, such as gain, return loss, reflection coefficient and amplifier stability, can be expressed by S-parameters easily. The S-parameters use matched loads, instead of open or short circuit conditions, so they are easier to be used in RF IC design. The design of EM simulation algorithms itself is a promising research field. Much work has been done [32]. The popular EM simulators are Momentum [33], CST [34], IE3D [31]. SPICE in RF IC design is used to perform circuit simulations using the S-parameter models provided by the EM simulator, such as S-parameter simulation. In addition, steady-state analysis is necessary for RF IC analysis, such as harmonic balance simulation.

- Design automation tools
  Analog EDA tools can be mainly classified into system design tools, topology generation tools, analog cell optimization tools, yield optimization tools and combinations of them. Some examples are in the following.

  - System design tools
    System design tools focus on analog system-level design. Some of them also include the lower-level cell sizing. For example, the multi-objective bottom-up (MOBU) method [35] constructs the analog system from the Pareto front of the basic cells by multi-objective optimization methods.
  - Topology generation tools
    Topology generation tools aid designers in the task of topology selection, design and sizing. A typical example is the MOJITO system [36]. MOJITO defines a space of thousands of possible topologies via a hierarchical organized combination of analog cells. The search algorithm traverses the design space and multi-objective optimization is performed.
  - Analog cell optimization tools
    Analog cell optimization tools size the analog building block with a predefined topology. Typical examples are [37, 38]. The tool optimizes the desired goal (e.g., power consumption) when satisfying the design specifications provided by the designer.
  - Yield optimization tools

High-yield design is extremely important in recent years. There are tools to enhance the yield for an available design by tuning the design parameters. More advanced tools start from sizing and consider yield in the sizing process to obtain a circuit optimized in both the performances and the yield. The latter case is also called yield / variation-aware analog IC sizing. Wicked [39] is a typical tool achieving this task.

– Combination tools
   Combination tools can finish more than one of the above tasks. For example, the MOJITO-N tool [40] can generate a circuit topology considering variation-aware sizing.

The design automation tools for RF ICs are still in their infancy. Although there are a few tools, their functions are quite limited. One kind of tools is for low-GHz (often less than 10 GHz) RF IC synthesis [41]. Such tools rely on equivalent circuit models for passive components at low GHz frequencies. Moreover, a layout template defining the locations of each component is often required, which restricts the design space. In other words, some good solutions may not be found because of the restrictions brought by the layout template. Such tools cannot solve mm-frequency RF IC synthesis problems. Another kind of tools, which is often in-house tools by RF IC designers, can solve mm-frequency RF IC optimization, but requires a good initial design by the designer. The function of the tool is to locally tune the parameters to enhance the performance. Strictly speaking, such tools can hardly be called design "automation" tools.

With this background, some people could think that: mathematically, RF IC optimization is the same as low-frequency analog IC optimization, and the simulation tools for RF IC are available. Then why not use the same "simulation in the loop" method to globally optimize RF ICs? So RF EDA is promising and easy to implement. This thought is partly right and partly wrong. The correct part is that the future of RF EDA is indeed very bright, because most of the important problems have not been solved yet. The incorrect part is that the "simulation in the loop" based optimization methods are not applicable to RF ICs. In Chap. 9, an example to design a mm-wave amplifier will be shown. By using the "simulation in the loop" method combined with powerful global optimization algorithms, the time consumption is 9 days. Although the optimized design is very good, such approach is not efficient enough to meet the time-to-market requirements. Hence, research is needed to speed up RF IC design automation, while maintaining the high optimization ability.

1.5 Summary

This introductory chapter has introduced the basic concepts and backgrounds of computation intelligence and analog and RF IC design automation. The development and challenges we have met in both fields have been reviewed. The rest of this book will elaborate the algorithm frameworks and the practical algorithms for the solution of the problems presented in this chapter.

References

1. Russell S, Norvig P, Canny J, Malik J, Edwards D (1995) Artificial intelligence: a modern approach. Prentice hall Englewood Cliffs, New Jersey
2. Eiben A, Smith J (2003) Introduction to evolutionary computing. Springer Verlag, Berlin
3. Dreslinski R, Wieckowski M, Blaauw D, Sylvester D, Mudge T (2010) Near-threshold computing: reclaiming moore's law through energy efficient integrated circuits. Proc IEEE 98(2):253–266
4. ITRS (Sept. 2011) ITRS report. http://www.itrs.net/
5. Liu B, Wang Y, Yu Z, Liu L, Li M, Wang Z, Lu J, Fernández F (2009d) Analog circuit optimization system based on hybrid evolutionary algorithms. Integr VLSI J 42(2):137–148
6. Gielen G, Eeckelaert T, Martens E, McConaghy T (2007) Automated synthesis of complex analog circuits. In: Proceedings of 18th european conference on circuit theory and design, pp 20–23
7. McConaghy T, Palmers P, Gao P, Steyaert M, Gielen G (2009a) Variation-aware analog structural synthesis: a computational intelligence approach. Springer Verlag, Berlin
8. Liu B, Fernández F, Gielen G (2011a) Efficient and accurate statistical analog yield optimization and variation-aware circuit sizing based on computational intelligence techniques. IEEE Trans Comput Aided Des Integr Circ Syst 30(6):793–805
9. Niknejad A, Hashemi H (2008) mm-Wave silicon technology: 60GHz and beyond. Springer Verlag, New York
10. Choi K, Allstot D (2006) Parasitic-aware design and optimization of a CMOS RF power amplifier. IEEE Trans Circ Syst I Regul Pap 53(1):16–25
11. Tulunay G, Balkir S (2008) A synthesis tool for CMOS RF low-noise amplifiers. IEEE Trans Comput Aided Des Integr Circ Syst 27(5):977–982
12. Ramos J, Francken K, Gielen G, Steyaert M (2005) An efficient, fully parasitic-aware power amplifier design optimization tool. IEEE Trans Circ Syst I Regul Pap 52(8):1526–1534
13. Balanis C (1982) Antenna theory: analysis and design. Wiley, New York
14. Poian M, Poles S, Bernasconi F, Leroux E, Steffé W, Zolesi M (2008) Multi-objective optimization for antenna design. In: Proceedings of IEEE international conference on microwaves, communications, antennas and electronic systems, pp 1–9
15. Yeung S, Man K, Luk K, Chan C (2008) A trapeizform U-slot folded patch feed antenna design optimized with jumping genes evolutionary algorithm. IEEE Trans Antennas Propag 56(2):571–577
16. Mezura-Montes E (2009) Constraint-handling in evolutionary optimization. Springer Verlag, Berlin
17. Hart W, Krasnogor N, Smith J (2005) Recent advances in memetic algorithms. Springer Verlag, Berlin
18. Fogel D (2006) Evolutionary computation: toward a new philosophy of machine intelligence. Wiley-IEEE Press, Piscataway
19. Coello C, Lamont G, Veldhuizen D (2007) Evolutionary algorithms for solving multi-objective problems. Springer-Verlag, New York
20. Price K, Storn R, Lampinen J (2005) Differential evolution: a practical approach to global optimization. Springer-Verlag, New York
21. Poli R, Kennedy J, Blackwell T (2007) Particle swarm optimization. Swarm Intell 1(1):33–57
22. Dorigo M, Birattari B, Stützle T (2006) Ant colony optimization. IEEE Comput Intell Mag 1(4):28–39
23. Ross T (1995) Fuzzy logic with engineering applications. Wiley, New York
24. Zadeh L (1965) Fuzzy sets. Inf control 8(3):338–353
25. Liu B (2002) Theory and practice of uncertain programming. Physica Verlag, Heidelberg
26. Eiben A, Bäck T (1997) Empirical investigation of multiparent recombination operators in evolution strategies. Evol Comput 5(3):347–365

27. Gielen G, McConaghy T, Eeckelaert T (2005) Performance space modeling for hierarchical synthesis of analog integrated circuits. In: Proceedings of the 42nd annual design automation conference, pp 881–886
28. Liu B, Deferm N, Zhao D, Reynaert P, Gielen G (2012b) An efficient high-frequency linear RF amplifier synthesis method based on evolutionary computation and machine learning techniques. IEEE Trans Comput Aided Des Integr Circ Syst 31(7):981–993
29. Synopsys (2013) HSPICE homepage. http://www.synopsys.com/Tools/Verification/ AMSVerification/CircuitSimulation/HSPICE/Pages/default.aspx
30. Cadence (2013) cadence design system homepage. http://www.cadence.com/us/pages/default. aspx
31. Mentor-Graphics (2013) Mentor graphics homepage. http://www.mentor.com/
32. Yu W (2009) Electromagnetic simulation techniques based on the FDTD method. Wiley, New York
33. Agilent (2013) Agilent technology homepage. http://www.home.agilent.com/
34. CST (2013) CST computer simulation technology homepage. http://www.cst.com/
35. Eeckelaert T, McConaghy T, Gielen G (2005) Efficient multiobjective synthesis of analog circuits using hierarchical pareto-optimal performance hypersurfaces. In: Proceedings of the conference on design, automation and test, pp 1070–1075
36. McConaghy T, Palmers P, Gielen G, Steyaert M (2007) Simultaneous multi-topology multi-objective sizing across thousands of analog circuit topologies. In: Proceedings of the 44th design automation conference, pp 944–947
37. Medeiro F, Rodríguez-Macías R, Fernández F, Domínguez-Castro R, Huertas J, Rodríguez-Vázquez A (1994b) Global design of analog cells using statistical optimization techniques. Analog Integr Circ Sig Process 6(3):179–195
38. Medeiro F, Fernández F, Dominguez-Castro R, Rodriguez-Vazquez A (1994a) A statistical optimization-based approach for automated sizing of analog cells. In: Proceedings of the IEEE/ACM international conference on Computer-aided design, pp 594–597
39. MunEDA (2013) MunEDA homepage. http://www.muneda.com/index.php
40. McConaghy T, Palmers P, Gielen G, Steyaert M (2008) Genetic programming with reuse of known designs for industrially scalable, novel circuit design, Chap. 10. Genetic Programming Theory and Practice V. Springer, pp 159–184
41. Allstot D, Choi K, Park J (2003) Parasitic-aware optimization of CMOS RF circuits. Springer, New York

Chapter 2
Fundamentals of Optimization Techniques in Analog IC Sizing

Chapters 2, 3, 4 concentrate on high-performance analog integrated circuit sizing under nominal conditions. This is the foundation of advanced topics in the following chapters of this book, i.e., variation-aware analog IC sizing and electromagnetic (EM)-simulation based mm-wave integrated circuit and antenna synthesis. This chapter defines the problem, introduces some widely used evolutionary algorithms and basic constraint handling approaches.

This chapter is organized as follows. Section 2.1 introduces the problem. Section 2.2 reviews existing approaches to analog circuit sizing. Section 2.3 introduces the general implementation of an EA and provides a detailed introduction to the differential evolution (DE) algorithm, which is used as the typical search mechanism throughout this book. Section 2.4 introduces two basic but widely used constraint handling techniques. Section 2.5 introduces two widely used multi-objective evolutionary algorithms (MOEAs). Two examples are shown in Sect. 2.6. Section 2.7 provides the summary of this chapter.

2.1 Analog IC Sizing: Introduction and Problem Definition

Nowadays, VLSI technology progresses towards the integration of mixed analog-digital circuits as a complete system-on-a-chip. Although the analog part is a small fraction of the entire circuit, it is much more difficult to design due to the complex and knowledge-intensive nature of analog circuits. Without an automated synthesis or sizing methodology, analog circuit design suffers from long design times, high design complexity, high cost and requires highly skilled designers. Consequently, automated synthesis methodologies for analog circuits have received much attention. The analog circuit design procedure consists of topological-level design and parameter-level design (also called circuit sizing) [1, 2]. This book concentrates on the latter, aiming at parameter selection and optimization to improve the performances for a given circuit topology. We assume that the designer provides the circuit topology.

B. Liu et al., *Automated Design of Analog and High-frequency Circuits*,
Studies in Computational Intelligence 501, DOI: 10.1007/978-3-642-39162-0_2,
© Springer-Verlag Berlin Heidelberg 2014

There are two main purposes of an analog IC sizing system: first, to replace exploration of tedious and ad-hoc manual trade-offs by automatic design of parameters; secondly, to solve problems that are hard to design by hand. Accuracy, ease of use, generality, robustness, and acceptable run-time are necessary for a circuit synthesis solution to gain acceptance [3]. Other than those requirements, the ability to deal with complex problems, closely meeting the designer's requirements even for highly constrained problems, and the ability to achieve highly optimized results are significant objectives of a high-performance analog IC sizing system. Many parameter-level design strategies, methods, and tools have been published in recent years [1–22], and some have even reached commercialization. We will review them in Sect. 2.2.

Most analog circuit sizing problems can naturally be expressed as a single- or multi-objective constrained optimization problem. We first consider single-objective cases. The problem can be defined as the minimization[1] of an objective, e.g., power consumption, usually subject to some constraints, e.g., DC gain larger than a certain value. Mathematically, this can be formulated as follows:

$$
\begin{aligned}
&\text{minimize}_x \, f(x) \\
&\text{s.t. } g(x) \geq 0 \\
&\quad h(x) = 0 \\
&\quad x \in [X_L, X_H]
\end{aligned}
\tag{2.1}
$$

In this equation, the objective function $f(x)$ is the performance function to be minimized. $h(x)$ are the equality constraints. In analog circuit design, the equality constraints mainly refer to Kirchhoff's current law (KCL) and Kirchhoff's voltage law (KVL) equations. Vector x corresponds to the design variables, and X_L and X_H are their lower and upper bounds, respectively. The vector $g(x) \geq 0$ corresponds to user-defined inequality constraints. An example is provided as follows: given the example circuit topology shown in Fig. 2.1, the sizing problem can be defined as (2.2). The objective of the sizing system is to determine the sizing and biasing of all devices (transistors, capacitors, etc) such that the power consumption is the smallest possible while satisfying all the constraints mentioned in (2.2).

$$
\begin{aligned}
&\text{minimize power} \\
&\text{s.t. DC gain} \geq 80\,\text{dB} \\
&\quad \text{GBW} \geq 2\,\text{MHz} \\
&\quad \text{Phase Margin} \geq 50° \\
&\quad \text{Slew Rate} \geq 1.5\,\text{V}/\mu\text{s}
\end{aligned}
\tag{2.2}
$$

For multi-objective analog circuit sizing, we simultaneously optimize more than one performance and get the approximate Pareto-optimal front. By multi-objective sizing, the trade-offs and sensitivity analysis between the different objectives can be

---

[1] The maximization of a design objective can easily be transformed into a minimization problem by just inverting its sign.

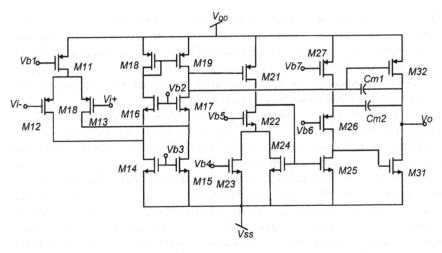

Fig. 2.1 A CMOS three-stage amplifier

explored. In multi-objective sizing, performances can either be set as objectives or constraints. An objective means that the designer is interested in its trade-offs with all other objectives in the whole performance space, which contains much information; while a constraint usually means that the performance needs to be larger than or smaller than a certain value, and no trade-off information is considered. Usually, only a part of the important performances are considered as objectives, and the others can be set as constraints. The reason is that the designer is often not interested in the trade-offs among all performances. Moreover, there are typically other design constraints, such as restricting the transistors to operate in the saturation region, which are essentially defined as functional constraints. An example is shown in (2.3).

$$
\begin{aligned}
&\text{minimize power}\\
&\text{minimize area}\\
&\text{s.t. DC gain} \geq 80\,\text{dB}\\
&\quad\text{GBW} \geq 2\,\text{MHz}\\
&\quad\text{Phase Margin} \geq 50^\circ\\
&\quad\text{Slew Rate} \geq 1.5\,\text{V}/\mu\text{s}
\end{aligned}
\tag{2.3}
$$

2.2 Review of Analog IC Sizing Approaches

Analog integrated circuit sizing can be carried out by the following two different approaches: knowledge-based and optimization-based [23]. The basic idea of knowledge-based synthesis is to formulate design equations in such a way that given the performance characteristics the design parameters can be directly calculated

[4–6]. In these tools, the quality of the solutions in terms of both accuracy and robustness are often not acceptable in complex circuits and modern technologies since the very concept of knowledge-based sizing forces the design equations to be simple [4]. Other drawbacks are the large preparatory time/effort required to develop design plans or equations, the difficulty in using them in a different technology, and the limitation to a limited set of circuit topologies [5].

In optimization-based synthesis, the problem is translated into a function minimization problem, which can be solved through numerical methods, such as (2.2). This kind of method is widely accepted [23]. Essentially, it is based on the introduction of a performance evaluator within an iterative optimization loop [23]. The system is called equation-based when the performance evaluator is based on equations capturing the behavior of a circuit topology [7–10]. However, creating the equations often consumes much more time than manually designing the circuit. In addition, the simplifications required in the closed-form analytical equations cause low accuracy and incompleteness. On the contrary, simulation-based methods do not rely on analytical equations but on SPICE-like simulations to evaluate the circuit performances in the optimization process (the values of $f(x)$ and $g(x)$ in (2.1) are derived by the electrical simulator), which results in superior accuracy, generality, and ease of use [12–14]. Therefore, the simulation-based optimization method is the main focus in this book. Through the link between an available circuit simulator (e.g., HSPICE [24]) and the environment of programming the optimization algorithm (e.g., MATLAB, C++ environment), the candidate parameter values are transmitted from the optimization system to the simulation engine, and the circuit performances obtained by the electrical simulator are returned to the optimization system. The penalty to pay is a relatively long computation time (compared to other methods), although, due to the fast speed of the circuit simulation software and computers nowadays, the computational time of a full sizing can be kept within very acceptable limits, normally from minutes to tens of minutes for a circuit up to tens of transistors.

Techniques for analog circuit optimization that have appeared in literature can broadly be classified into two main categories: deterministic optimization algorithms and stochastic search algorithms (evolutionary computation algorithms, simulated annealing, etc). The traditional deterministic optimization methods include Newton methods, Levenberg-Marquardt method, etc. These techniques are available in some commercial electrical simulators [24]. The drawbacks of deterministic optimization algorithms are mainly the following three aspects: (1) they require a good starting point; (2) an unsatisfactory local minimum may be reached in many cases; (3) they often require continuity and differentiability of the objective function. Some researchers have tried to address these difficulties, such as [15], where a method to determine the initial point is presented. Another approach is the application of geometric programming methods, which guarantee the convergence to a global minimum [10]. However, they require a special formulation of the design equations, which make them share many of the disadvantages of equation-based methods. Research efforts on stochastic search algorithms, especially evolutionary computation (EC) algorithms (genetic algorithms, differential evolution, genetic programming, etc) have begun to appear in literature in recent years [1, 16–22]. Due to the ability and efficiency to find a satisfactory solution in a reasonable CPU time, genetic algorithm (GA) has been employed as optimization routines for analog circuits in both industry and academia. For problems with practical design constraints, most of the reported approaches use the penalty function method to handle the constraints [7–11, 23].

2.3 Implementation of Evolutionary Algorithms

2.3.1 Overview of the Implementation of an EA

In Chap. 1, the basic concepts of evolutionary computation have been introduced using the example of the genetic algorithm. This subsection concentrates on the implementation of an EA. An EA program often follows the following procedure:

Step 1: Initialize the population of individuals by random generation.
Step 2: Evaluate the fitness of each individual in the initial population.
Step 3: Evolution process until the termination condition is met:

> **Step 3.1:** Select the parent individuals for reproduction.
> **Step 3.2:** Generate offsprings (child population) based on the parent individuals through crossover and mutation.
> **Step 3.3:** Evaluate the fitness of each individual in the child population.
> **Step 3.4:** Update the population by replacing less fitting individuals by individuals with good fitness.

Different EAs can differ from the representation of solutions to the search operators. For the representation of solutions, strings of numbers (e.g., binary), real numbers and computer program structure can be used, and the typical corresponding EAs are canonical genetic algorithm, evolution strategy and genetic programming, respectively. In terms of the search operators, there are a number of mutation, crossover and selection operators. Note that appropriate mutation, crossover and selection operators must be combined to achieve a good search performance.

The targeted problems in this book, the design automation of analog and RF IC, are numerical optimization problems. In the following, we will introduce the differential evolution (DE) algorithm [25, 26] as an example of the implementation of EA. DE is based on the vector differences and is especially suitable for numerical optimization problems. For many continuous global optimization problems, DE is often the first choice and is used as the search mechanism throughout this book. Besides DE, other promising CI approaches, real-coded genetic algorithm [27, 28], particle swarm optimization [29] and evolution strategy [30, 31] have good potential for the targeted problems, such as [32].

2.3.2 Differential Evolution

In general, the goals in this book are to optimize certain properties of a system by pertinently choosing the system parameters. Moreover, because of using simulation to evaluate the candidate's performances, the problems are often black-box optimization problems. They may have the properties of being nonlinear, non-convex and non-differentiable. As said above, among evolutionary algorithms, DE [25, 26] is recognized as a very effective evolutionary search engine for global optimization over continuous spaces.

Users generally demand that a practical optimization technique fulfills the following requirements [25] (minimization is considered):

(1) Ability to handle non-differentiable (gradients are not available or difficult to be calculated), nonlinear and multimodal cost functions (more than one or even numerous local optima exist).
(2) Parallelizability to cope with computation-intensive cost functions.
(3) Ease of use, i.e., few control variables to steer the minimization. These variables should also be robust and easy to choose.
(4) Good convergence properties, i.e., consistent convergence to the global minimum in consecutive independent trials.

As explained in the following, DE was designed to fulfill all of the above requirements. For the first two requirements, they are common to all EAs, so does DE. In order to satisfy requirement (3), DE borrows the idea from the Nelder and Mead algorithm [33] of employing information from within the vector population to alter the search space. DE's self-organizing scheme takes the difference vector of two randomly chosen population vectors to perturb an existing vector. Extensive testing under various conditions show the high performance of DE for complex benchmark problems [25, 26]. For requirement (4), although theoretical description of the convergence properties are available for many approaches, only extensive testing under various conditions can show whether an optimization method can fulfill its promises. DE scores very well in this regard for complex benchmark problems [25, 26].

In the following, the DE operations are described in detail.

Differential Evolution (DE) is a parallel direct search method which utilizes NP d-dimensional parameter vectors:

$$x_i(t) = [x_{i,1}, x_{i,2}, \ldots, x_{i,d}], \quad i = 1, 2, \ldots, NP \tag{2.4}$$

as a population for each iteration t. NP is the size of the population and does not change during the minimization process. The initial vector population is chosen randomly and should cover the entire parameter space. As a rule, we will assume a uniform probability distribution for all random decisions unless otherwise stated. In case a preliminary solution is available, the initial population might be generated by adding normally distributed random deviations to the nominal solution. DE generates new parameter vectors by adding the weighted difference between two population vectors to a third vector. Let this operation be called mutation. The mutated vector's parameters are then mixed with the parameters of another predetermined vector (the corresponding vector generated in the last iteration), the target vector, to yield the so-called trial vector. Parameter mixing is often referred to as "crossover" in the EC community and will be explained later in more detail. If the trial vector yields a lower cost function value than the target vector, the trial vector replaces the target vector in the following generation. This last operation is called selection. Each population vector has to serve once as the target vector so that NP competitions take place in one generation (iteration).

More specifically DE's basic strategy can be described as follows:

A. Mutation

For each target vector $x_i(t); i = 1, 2, \ldots, NP$, a mutant vector is generated according to

$$V_i(t+1) = x_{r1}(t) + F(x_{r2}(t) - x_{r3}(t)) \qquad (2.5)$$

where random indexes $r1, r2, r3 \in 1, 2, \ldots, NP$, are integer, mutually different numbers and $F > 0$. The randomly chosen integers $r1, r2$ and $r3$ are also chosen to be different from the running index i, so that NP must be greater or equal to four to allow for this condition. F is a real and constant number within $(0, 2]$ which controls the amplification of the differential variation $x_{r2}(t) - x_{r3}(t)$. The selection of F is detailed in [26]. For single-objective problems, using F between 0.8–1 is a common choice. For multi-objective optimization using DE as the search engine, this depends on the specific search mechanism, which will be illustrated in the following chapters. Figure 2.2 shows a two-dimensional example that illustrates the different vectors which play a part in the iteration of $V_i(t + 1)$.

B. Crossover

In order to increase the diversity of the perturbed parameter vectors, crossover is introduced. To this end, the trial vector:

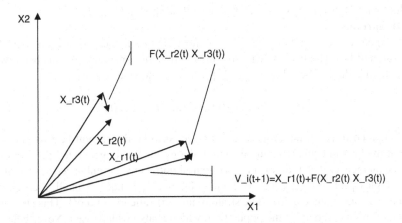

Fig. 2.2 An example of the process for generating $V_i(t + 1)$ (two-dimensional)

Fig. 2.3 Illustration of the crossover process for $D = 7$ parameters (from [25])

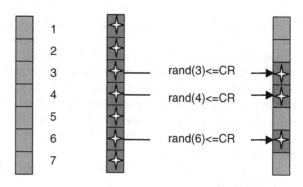

$$U_i(t + 1) = [u_{i,1}(t + 1), u_{i,2}(t + 1), \ldots, u_{i,d}(t + 1)], \quad i = 1, 2, \ldots, NP \quad (2.6)$$

is formed, where

$$u_{i,j}(t + 1) = \begin{cases} v_{i,j}(t + 1), & \text{if } (rand(i, j) \le CR) | j = randn(i) \\ x_{i,j}(t), & \text{otherwise} \end{cases} \quad (2.7)$$

In (2.7), $rand(i, j)$ is the j-th evaluation of a uniform random number generator with outcome within [0, 1]. CR is the crossover constant, within [0, 1], which has to be determined by the user. $randn(i)$ is a randomly chosen index $1, 2, \ldots, d$ which ensures that $U_i(t + 1)$ gets at least one parameter from $V_i(t + 1)$. Figure 2.3 gives an example of the crossover mechanism for 7-dimensional vectors.

C. Selection

To decide whether or not the trial vector $U_i(t + 1)$ should become a member of generation $t + 1$, it is compared to the target vector $x_i(t)$ using the greedy criterion. If vector $u_i(t + 1)$ yields a smaller fitness function value than $x_i(t)$, then $x_i(t)$ is set to $u_i(t + 1)$; otherwise, the old value $x_i(t)$ is retained. DE follows a standard evolutionary algorithm flow, which has been described in Chap. 1.

DE variants

It can be seen that mutation (2.5) is the most important operation which controls the search of DE. There are at least six kinds of mutation operators. In this book, we normally use the DE/best/1 strategy [25], which is shown as follows:

$$V_i(t + 1) = x_{best}(t) + F(x_{r2}(t) - x_{r3}(t)) \quad (2.8)$$

Compared to (2.5), the base vector is selected to be the current best candidate, so that the best information can be shared among the newly generated population. The advantage is that the convergence can significantly be enhanced, especially for complex problems [26, 34]. If the population size is not too small, the risk of premature convergence is low. The strategy in (2.5) called DE/rand/1. This strategy maintains high diversity of the population, which is also widely used. Nevertheless, its convergence speed is often slower.

2.4 Basics of Constraint Handling Techniques

The constraint handling problem is very important in many real-world optimization problems, including analog circuit sizing. For high-performance (with tough specifications) circuits, the optimization problems are always highly constrained. However, the evolutionary algorithms, serving as the optimization engine, are constraint blind. Hence, constraint handling technique is a hot topic in the EC field. This section will introduce two widely applied constraint handling methods. They are often used in current literatures for analog IC sizing.

2.4.1 Static Penalty Functions

Most of the reported synthesis methods use static penalty functions to handle constraints [23, 35]. In these methods, the constrained optimization problem is transformed into an unconstrained one by minimizing the following function (a minimization problem is considered):

$$f'(x) = f(x) + \sum_{i=1}^{i=n} w_i < g_i(x) > \tag{2.9}$$

where the parameters w_i are the penalty weighting coefficients. "Static" means the penalty weighting coefficients are decided beforehand and are not changed during the optimization process. $\langle g_i(x) \rangle$ returns the absolute value of $g_i(x)$ if its negative, and zero otherwise, considering the constraints of $g_i(x) \geq 0, i = 1, 2, \ldots, n$. $f(x)$ is the objective function.

Clearly, the advantage of the static penalty function-based method is its simplicity. Only the fitness function is properly formulated and all the evolutionary operators do not need any adaptation or revision. On the other hand, the outcome of approaches based on static penalty function techniques is sensitive to the values chosen for the penalty coefficients, but the determination of proper penalty coefficients is a tough work. Small values of the penalty coefficients drive the search outside the feasible region and often produce infeasible solutions, whereas imposing very severe penalties makes it difficult to drive the population to the optimum. Although several penalty strategies have been developed, there has been no general rule for designing penalty coefficients. As the experiments in Chap. 3 will demonstrate, the methodology based on the combination of genetic algorithms and penalty functions cannot successfully solve design problems with many or severe constraints.

2.4.2 Selection-Based Constraint Handling Method

A constraint handling algorithm based on tournament selection for genetic algorithms that has proven to be effective was proposed by [36]. Prior to this, identical separation of feasible and infeasible solutions had been proposed in combination with simulated annealing by analog IC sizing researchers [13, 37]. Given two candidates in the population, there may be, at most, three situations:

(1) Both solutions are feasible;
(2) Both solutions are infeasible;
(3) One solution is feasible, but the other is not.

Accordingly, the selection rules are:

(1) Given two feasible solutions, select the one with the better objective function value;
(2) Given two infeasible solutions, select the solution with the smaller constraint violation;
(3) If one solution is feasible and the other is not, select the feasible solution.

The most important advantage is that this method needs no penalty coefficients, so the problems of penalty function-based algorithms are overcome. The drawback is that some candidates with very good performances may be missing in the search process because of loss of diversity. This deficiency will occur in problems with disconnected feasible regions in which cases the EA may be stuck within one of the feasible regions and never get to explore the others [38]. To illustrate this phenomenon, let us consider the performance space in Fig. 2.4. Assume that the regions in grey represent the feasible regions (they are disconnected) and that the global optimum is in the left one. In the beginning, the evolution may follow a fast way to minimize the total violation of constraints without considering the objective function, causing some decision variable(s) swiftly go(es) into one of the feasible intervals and is / are restricted there. In this example, the population after some iterations may converge in the feasible region on the right, but the global optimal value for the objective function is in the feasible region on the left.

   Although these two methods have some drawbacks, they are often workable for analog circuit sizing problems with ordinary constraints. For highly constrained or problems with complex hyper-surfaces, which may appear in high-performance analog circuit sizing, advanced methods are necessary.

Fig. 2.4 Illustrative solution space

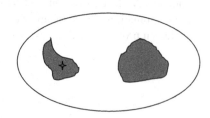

2.5 Multi-objective Analog Circuit Sizing

Multi-objective analog circuit sizing is based on multi-objective evolutionary algorithms (MOEAs). The main difference between multi-objective optimization and single-objective optimization is the fitness assignment. For single-objective optimization, the optimality is determined by a single function value, but for multi-objective optimization, not only the optimality should be determined by multiple objective functions, but also the distribution of the solutions in the approximated Pareto front (PF) is important. MOEAs can be generally classified into non-dominate sorting-based methods and decomposition-based methods. This section will introduce non-dominated sorting genetic algorithm-II (NSGA-II) [39] and multi-objective evolutionary algorithm based on decomposition (MOEA/D) [40], which are widely used in multi-objective sizing of analog circuits.

Before introducing MOEAs, two essential concepts for Pareto-optimal multi-objective optimization are described.

- Dominance
  Considering that there are two objectives $f_1(x)$ and $f_2(x)$, and both of them are for minimization, let x and x' be two candidate solutions for the multi-objective optimization problem. x is said to dominate x' if and only if $f_1(x) \leq f_1(x')$, $f_2(x) \leq f_2(x')$, and at least one of these two inequalities is strict. A solution x^* is Pareto-optimal if there is no other solution that dominates it. The set of all the Pareto-optimal solutions is called the Pareto set (PS) and the image of PS in the objective space (i.e., $f_1 - f_2$ space) is the Pareto front (PF). All the Pareto-optimal solutions are considered as equally good if there is no specific preference.
- Diversity of the approximated PF
  A decision maker often wants to have an approximate PF for gaining more understanding of the problem to make his / her final decision. Note that in most MOEAs, the PF is approximated by a number of points (Pareto-optimal solutions). Therefore, it is desirable that the generated Pareto-optimal points spread evenly in the approximated PF (high diversity), instead of clustered to a/several small part(s) of the PF (low diversity), where no information can be gained from the blanks in the approximated curves or hyper-surfaces.

2.5.1 NSGA-II

NSGA-II uses non-dominated sorting for fitness assignments. As has been said, in multi-objective sizing, not only good convergence is the optimization goal, but also the diversity of the points in the PF is important. Thus, environmental selection, that considers both dominance and distance between individuals, is necessary [41]. The truncation method is the approach for environmental selection very commonly used in recent years for multi-objective optimization problems with 2–3 objectives [41] and is also used in NSGA-II. A typical truncation method proceeds as follows:

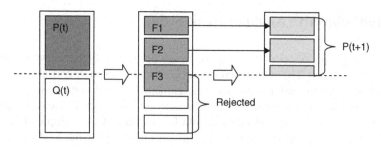

Fig. 2.5 Truncation method used in NSGA-II from [39]

(1) Collect all the candidates that compete for the next generation and rank them on dominance.
(2) If the number of non-dominated solutions exceeds the size of the population, then the non-dominated solutions are ranked by distance. The solutions with ranks larger than the population size are cut off.
(3) If the number of non-dominated solutions is less than the population size, then fill in the available places with dominated solutions according to their quality on dominance and distribution.

Figure 2.5 shows the truncation method, where P means the parent population, Q means the offspring population and F means the number of fronts.

The front number assigned to each individual is ranked by the level of other individuals that dominate it. The pseudocode of the fast non-dominate sorting is in Fig. 2.6 [39]. In order to make the algorithm evolve towards a uniformly spread out PF, crowding distance is used and is considered in the ranking. Crowding distance estimates the density of solutions surrounding a particular solution in the population. It calculates the average distance of two points on either side of this point along each of the objectives. For each objective, the population is firstly sorted. Then, the solutions with smallest and largest function values are assigned an infinite distance value. All other intermediate solutions are assigned a distance value equal to the absolute normalized difference in the function values of two adjacent solutions. The crowding-distance value is the sum of the normalized distance values of each objective.

The search engine of NSGA-II is a real coded GA. The crossover operator used is simulated binary crossover (SBX). The procedure of generating two children c_1 and c_2 from two parent solutions p_1 and p_2 is as follows:

1. Create a random number u between 0 and 1.
2. Calculate coefficient β:

$$\beta = \begin{cases} (2u)^{1/(\eta+1)} & \text{if } u \leq 0.5 \\ (\frac{1}{2(1-u)})^{1/(\eta+1)}, & otherwise \end{cases} \tag{2.10}$$

```
fast-non-dominated-sort(P)
for each p ∈ P
    S_p = ∅
    n_p = 0
    for each q ∈ P
        if (p ≺ q) then              If p dominates q
            S_p = S_p ∪ {q}          Add q to the set of solutions dominated by p
        else if (q ≺ p) then
            n_p = n_p + 1            Increment the domination counter of p
    if n_p = 0 then                  p belongs to the first front
        p_rank = 1
        F_1 = F_1 ∪ {p}
i = 1                                Initialize the front counter
while F_i ≠ ∅
    Q = ∅                            Used to store the members of the next front
    for each p ∈ F_i
        for each q ∈ S_p
            n_q = n_q - 1
            if n_q = 0 then          q belongs to the next front
                q_rank = i + 1
                Q = Q ∪ {q}
    i = i + 1
    F_i = Q
```

Fig. 2.6 Fast non-dominate sorting method in NSGA-II from [39]

3. The children solutions are

$$c_1 = 0.5[(1 - \beta)p_1 + (1 + \beta)p_2]$$
$$c_2 = 0.5[(1 + \beta)p_1 + (1 - \beta)p_2] \tag{2.11}$$

where the positive number η is the distribution index of the SBX operator. It can be seen that a small η would generate children solutions far away from the parent solutions, while a large η restricts children solutions to be near the parent solutions.

Essentially, the SBX operator has two properties:

• The difference between the offspring is in proportion to the parent solution.
• Near-parent solutions become mostly offspring rather than solutions distant from parents if η is properly selected.

The mutation operator used in NSGA-II is polynomial mutation. The probability distribution is a polynomial function. The shape of the probability distribution is directly controlled by an external parameter γ_m, and the distribution is not dynamically changed with iterations. If x_i is the value of the ith parameter selected for mutation with a probability p_m and the result of the mutation is the new value y_i obtained by a polynomial probability distribution $Pr(\delta) = 0.5(\gamma_m + 1)(1 - |\delta|)^{\gamma_m}$, then:

$$y_i = x_i + (x_i^H - x_i^L)\delta_i$$

$$\delta_i = \begin{cases} (2r_i)^{1/(\gamma_m+1)-1} & \text{if } r_i < 0.5 \\ 1 - |2(1 - r_i)|^{1/(\gamma_m+1)}, & \text{otherwise} \end{cases} \qquad (2.12)$$

where x_i^L and x_i^H are the lower and upper bound of x_i respectively and r_i is a random number in [0, 1].

2.5.2 MOEA/D

MOEA/D is a recently proposed MOEA [40], which outperforms NSGA-II for complex problems and problems with more objectives (e.g., 5). By using a (linear or nonlinear) weighted aggregation method, the approximation of the PF can be decomposed into N single objective optimization subproblems. MOEA/D defines neighbourhood relations among these subproblems based on the distances among their weight vectors. Each subproblem is optimized in MOEA/D by using information mainly from its neighbouring subproblems. Among several decomposition methods, the Tchebycheff approach is the most widely used one. More specifically, the scalar function is as follows:

$$\begin{aligned} \text{minimize} \quad & g^{te}(x|\lambda, z^*) = max_{1 \le i \le m}\{\lambda_i|f_i(x) - z_i^*|\} \\ s.t. \quad & x \in [a, b]^d \end{aligned} \qquad (2.13)$$

where $\lambda = \{\lambda_1, \lambda_2, \ldots, \lambda_m\}$ is a weight vector and $\sum_{i=1}^{m} \lambda_i = 1$. $[a, b]^d$ is the solution space and $z^* = \{z_1^*, \ldots, z_m^*\}$ is the reference point. In other words, z^* are the smallest values of each objective function. If N is reasonably large and $\lambda^1, \ldots, \lambda^N$ are properly selected, the optimal solutions to those scalar functions will provide a good approximation to the Pareto Set (PS)/PF. It is worth noting that as z^* is usually unknown before the search, the smallest values of each objective found during the search are often used to substitute them. There are a number of different variants of MOEA/D. The original MOEA/D uses the SBX and polynomial mutation introduced in the previous subsection. MOEA/D-DE was proposed in [42] and high performance is shown.

MOEA/D-DE is described as follows.[2]

Input:

    (1) an multi-objective optimization problem (MOP)
    (2) a stopping criterion
    (3) N: the number of sub-problems

---

[2] Note that in the algorithmic description for multi-objective optimization, to avoid the confusion of indices in a vector and the indices in a group of vectors, a superscript i indicates the ith individual of a group, and a subscript i indicates the ith element in a vector.

(4) T: the neighborhood size
(5) δ: the probability that parent solutions are selected from the neighborhood
(6) n_r: the maximum number of solutions replaced by a child solution
(7) CR: the crossover rate in DE
(8) $\widehat{F}$: the scaling factor in the DE mutation
(9) p_m: the probability to perform polynomial mutation
(10) λ: the weight vector (the generation method is in [40])

Output:

(1) Approximation to the PF
(2) Approximation to the PS

Procedure:

Step 1: Initialization

Step 1.1: Compute the Euclidean distances between the weight vectors and work out the T closest weight vectors to each weight vector. For $i = 1, \ldots, N$, set $B(i) = \{i_1, \ldots, i_T\}$. $\lambda^{i_1}, \ldots, \lambda^{i_T}$ are the T closest vectors to λ^i.
Step 1.2: Randomly generate an initial population $x^1, \ldots, x^N$. Calculate the fitness values of the population.
Step 1.3: Initialize $z = \{z_1, \ldots, z_m\}$, where $z_j = min_{1 \leq i \leq N} f_j(x^i)$.

Step 2: Update
For $i = 1, \ldots, N$,

Step 2.1: Selection of the mating pool:
Generate a random number "rand" which is uniformly distributed in the range [0,1]. Set

$$P = \begin{cases} B(i), & \text{if } rand < \delta \\ \{1, \ldots, N\}, & \text{otherwise} \end{cases} \tag{2.14}$$

Step 2.2: Reproduction:
Set $r_1 = i$ and randomly select two indexes r_2 and r_3 from P, and generate a new solution $\bar{y}$ by the DE mutation. Then, perform a polynomial mutation on $\bar{y}$ with probability p_m to produce a new solution y.
Step 2.3: Repair:
If an element of y is out of the bound of $[a, b]^d$, its value is reset to be a randomly selected value inside the boundary.
Step 2.4: Update of the reference point:
For $j = 1, \ldots, m$, if $z_j > f_j(y)$, set $z_j = f_j(y)$.
Step 2.5: Update of solutions:
Set $c = 0$ and then do the following:
(1) If $c = n_r$ or P is empty, go to Step 3. Otherwise, randomly pick an index j from P.
(2) If $g^{te}(y|\lambda^j, z) \leq g^{te}(x^j|\lambda^j, z)$, then set $x^j = y$, and $c = c + 1$.
(3) Remove j from P and go to 1).

Step 3: Stopping Criterion:

If the stopping criterion (e.g., a certain number of iterations) is satisfied, then stop the algorithm and output $\{x^1, \ldots, x^N\}$ and $\{f(x^1), \ldots, f(x^N)\}$. Otherwise, go to Step 2.

2.6 Analog Circuit Sizing Examples

This section shows the application of the optimization techniques mentioned above. According to the previous sections, we have several effective single- and multi-objective evolutionary algorithms (GA, DE, PSO, ES, NSGA-II, MOEA/D) and two constraint handling methods (static penalty function and selection-based methods). To compose a method for analog circuit sizing, we can combine methods from the two pools. For example, for single-objective analog circuit sizing problem, the method of GA with static penalty function methods can be used. For multi-objective optimization without constraints, MOEA/D-DE can be used. Note that in the framework of MOEA/D, it is difficult to directly cooperate with a constraint handling method. This issue will be further discussed in Chap. 6.

Two examples are provided in the following: the first one is an analog circuit sizing problem based on single-objective constrained optimization, and the other is based on multi-objective constrained optimization.

2.6.1 Folded-Cascode Amplifier

The example circuit selected in this section is a folded-cascode amplifier. The folded-cascode amplifier is shown in Fig. 2.7. The load capacitance cl is 5 pF. The technology used is a 0.18 μm CMOS process with 1.8 V power supply. Transistor lengths were allowed to vary between the minimum value allowed by the technological process, 0.18 μm, up to 10 μm. Transistor widths were changed between the minimum technology value, 0.24 μm, up to 1000 μm. The bias current was allowed to vary between 1 μA and 2.5 mA. Appropriate matching relations were imposed: $M1 \equiv M2, M3 \equiv M4 \equiv Mbp, Mbn \equiv M5, M6 \equiv M7, M8 \equiv M9, M10 \equiv M11$. Therefore, the number of independent design parameters is 13.

2.6.2 Single-Objective Constrained Optimization

For the automated sizing of the above folded-cascode amplifier, the optimization goal is to minimize the power consumption with constraints on DC gain, gain-bandwidth product (GBW), phase margin, output swing, slew rate, area consumption and the functional constraints that all the transistors should be in the saturation region.

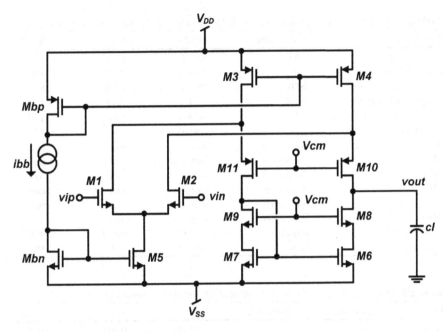

Fig. 2.7 CMOS Folded-cascode amplifier

The following methods are implemented: (1) the genetic algorithm, combined with penalty functions to handle constraints (GAPF); (2) the differential evolution algorithm combined with the same penalty-based method to handle constraints (DEPF); (3) an algorithm that combines the selection-based method for constrained optimization proposed by [36] and differential evolution (denoted as SBDE) [43]. For methods using DE, the DE/best/1 strategy is used. The inputs to the design methodology are a SPICE netlist file containing the topology and user-defined performance simulations. All examples have been run on a workstation with a Dual Xeon Quad processor at 3 GHz with 33 GB of RAM memory and Linux operating system.

Relatively soft performance constraints are used first. The design objective (power minimization) and constraints are shown in Table 2.1 together with the results provided by the three methods, GAPF, DEPF and SBDE. The penalty coefficients in the GAPF and DEPF algorithms were manually improved through five runs. At each new run, the penalty coefficients were updated, trying to increase the relative importance of the constraints not met in the previous run. Table 2.1 shows the best results among the five runs. It can be seen that all the algorithms met the design specifications.

Being a stochastic search process, a statistical study is needed to test the robustness of the different algorithms. Therefore, we executed 20 runs of each algorithm starting from 20 different initializations. For the GAPF and DEPF algorithms a reasonable choice of the penalty parameters was used for this statistical study: all constraints were normalized with respect to the specified value and equal weights were assigned to all of them. Table 2.2 shows the results for the different algorithms. The first two

Table 2.1 Design specifications and results of GAPF, DEPF and SBDE (not severe constraints)

Specifications	Constraints	GAPF	DEPF	SBDE
DC gain (dB)	≥ 55	62.14	55.03	56.87
GBW (MHz)	≥ 2	2.05	2.12	2.03
Phase margin (°)	≥ 50	88.25	83.58	82.09
Output swing (V)	≥ 1.2	1.31	1.39	1.28
Slew rate (V/μs)	≥ 1	1.13	1.00	1.02
Area (μm²)	≤ 225	196.95	150.53	142.16
*M*1	Saturation	met	met	met
*M*4	Saturation	met	met	met
*M*5	Saturation	met	met	met
*M*6	Saturation	met	met	met
*M*8	Saturation	met	met	met
*M*10	Saturation	met	met	met
Power (mW)	objective	0.028	0.024	0.023
Time (s)	N.A.	165	151	163

Table 2.2 Statistical results with the different methods (not severe constraints)

Item	Feasible	Infeasible	Best	Worst	Average
GAPF	20	0	0.0219	0.0473	0.0283
DEPF	20	0	0.0221	0.0282	0.0234
SBDE	20	0	0.0216	0.0305	0.0236

columns show the number of feasible and infeasible solutions found in the 20 runs. It can be seen that all methods were able to find a feasible solution. The following three columns in Table 2.2 show the best, worst and average value of the power consumption obtained in those 20 runs.

A set of more severe design constrains is shown in Table 2.3, together with a typical result provided by all algorithms. It can be observed that neither the GAPF algorithm nor the DEPF algorithm is able to satisfy the design constraints, even though five different sets of penalty coefficients were tried.

2.6.3 Multi-objective Optimization

For the same folded-cascode amplifier, the two optimization objectives are power and area, and the constraints include specifications on DC gain, GBW, phase margin, output swing and the functional constraints of all the transistors should be in the saturation region. This is a constrained multi-objective optimization problem. NSGA-II is used for multi-objective optimization. To handle constraints, the selection-based methods can be used to update the truncation method in NSGA-II. Feasibility (the

Table 2.3 Design specifications and results of GAPF, DEPF and SBDE (severe constraints)

Specifications	Constraints	GAPF	DEPF	SBDE
DC gain (dB)	≥ 60	61.89	60	60.06
GBW (MHz)	≥ 80	3.13	51.13	80.05
Phase margin (°)	≥ 60	79.59	78.84	74.74
Output swing (V)	≥ 1.25	1.39	1.32	1.26
Slew rate (V/μs)	≥ 60	1.56	33.97	60.00
Area (μm²)	≤ 440	330.22	320.20	431.73
*M*1	Saturation	met	met	met
*M*4	Saturation	met	met	met
*M*5	Saturation	met	met	met
*M*6	Saturation	met	met	met
*M*8	Saturation	met	met	met
*M*10	Saturation	met	met	met
Power (mW)	objective	0.03	0.74	1.31
Time (s)	N.A.	173	161	209

Table 2.4 Experimental results with two sets of constraints

Specifications	Constraints	Average	Constraints	Average
DC gain (dB)	≥ 60	60.05	≥ 55	55.05
GBW (MHz)	≥ 50	50.03	≥ 40	40.02
Phase margin (°)	≥ 60	82.60	≥ 60	85.72
Output swing (V)	≥ 1.2	1.25	≥ 1	1.23
Slew rate (V/μs)	≥ 30	34.00	≥ 20	22.86
*dm*1	≥ 1	17.64	≥ 1	17.58
*dm*3	≥ 1	1.81	≥ 1	1.67
*dm*5	≥ 1	3.52	≥ 1	4.24
*dm*7	≥ 1	2.42	≥ 1	2.23
*dm*9	≥ 1	1.56	≥ 1	1.79
*dm*11	≥ 1	7.56	≥ 1	8.28
Time (s)	N.A.	1706	X	1672

sum of the violation of constraints) is assigned superiority, and the truncation method in NSGA-II is only used for feasible individuals.

Two experiments have been performed with the severe constraints indicated in the second column of Table 2.4 and a second experiment with the not severe constraints from the fourth column. One typical solution is plotted for them, respectively. The power versus area Pareto fronts are shown in Fig. 2.8. The high specifications (severe constraints) and low specifications (not severe constraints) and the average values of the 200 individuals in the typical solution on each set of specifications are shown in Table 2.4.

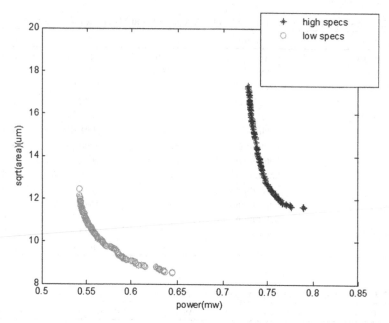

Fig. 2.8 Two typical Pareto fronts obtained for the two sets of constraints

2.7 Summary

This chapter has introduced the basics of automated sizing of analog circuits. It has tried to answer the following two questions: (1) How to use the evolutionary computation-based optimization techniques to solve practical problems in electronic design automation field; (2) How to design a workable analog circuit sizing method. The analog IC sizing problem is formulated and is connected with evolutionary computation techniques. After reviewing the previously published automatic analog circuit sizing approaches, the widely used and promising evolutionary algorithms for the targeted problem have been reviewed, among which, the differential evolution algorithm and its implementation were paid special attention. Constraint handling techniques and multi-objective evolutionary algorithms are also important in analog IC sizing and other real-world problems. Hence, two constraint handling techniques have been introduced. They are easy to implement and are effective to many problems, especially the selection-based methods. Two multi-objective optimization methods have then been introduced. NSGA-II is the most widely used method in multi-objective analog IC sizing and MOEA/D is a new state-of-the-art multi-objective optimization method, which is gradually becoming popular in the analog IC sizing field. At last, two practical examples have been shown to illustrate how to design an analog circuit sizing approach.

References

1. Nam D, Seo Y, Park L, Park C, Kim B (2001) Parameter optimization of an on-chip voltage reference circuit using evolutionary programming. IEEE Trans Evol Comput 5(4):414–421
2. Martens E, Gielen G (2008) Classification of analog synthesis tools based on their architecture selection mechanisms. Integr VLSI J 41(2):238–252
3. Krasnicki M, Phelps R, Hellums J, McClung M, Rutenbar R, Carley L (2001) ASF: a practical simulation-based methodology for the synthesis of custom analog circuits. In: Proceedings of IEEE/ACM international conference on computer aided design, pp 350–357
4. Degrauwe M, Nys O, Dijkstra E, Rijmenants J, Bitz S, Goffart L, Vittoz E, Cserveny S, Meixenberger C, Van Der Stappen G et al (1987) IDAC: an interactive design tool for analog CMOS circuits. IEEE J Solid-State Circuits 22(6):1106–1116
5. Harjani R, Rutenbar R, Carley L (1989) OASYS: a framework for analog circuit synthesis. IEEE Trans Comput Aided Des Integr Circuits Syst 8(12):1247–1266
6. Makris C, Toumazou C (1995) Analog IC design automation. ii. automated circuit correction by qualitative reasoning. IEEE Trans Comput Aided Des Integr Circuits Syst 14(2):239–254
7. Ochotta E, Rutenbar R, Carley L (1996) Synthesis of high-performance analog circuits in ASTRX/OBLX. IEEE Trans Comput Aided Des Integr Circuits Syst 15(3):273–294
8. Gielen G, Walscharts H, Sansen W (1990) Analog circuit design optimization based on symbolic simulation and simulated annealing. IEEE J Solid-State Circuits 25(3):707–713
9. Maulik P, Carley L, Allstot D (1993) Sizing of cell-level analog circuits using constrained optimization techniques. IEEE J Solid-State Circuits 28(3):233–241
10. Boyd S, Lee T et al (2001) Optimal design of a CMOS op-amp via geometric programming. IEEE Trans Comput Aided Des Integr Circuits Syst 20(1):1–21
11. Nye W, Riley D, Sangiovanni-Vincentelli A, Tits A (1988) DELIGHT. SPICE: an optimization-based system for the design of integrated circuits. IEEE Trans Comput Aided Des Integr Circuits Syst 7(4):501–519
12. Phelps R, Krasnicki M, Rutenbar R, Carley L, Hellums J (2000) Anaconda: simulation-based synthesis of analog circuits via stochastic pattern search. IEEE Trans Comput Aided Des Integr Circuits Syst 19(6):703–717
13. Medeiro F, Rodríguez-Macías R, Fernández F, Domínguez-Castro R, Huertas J, Rodríguez-Vázquez A (1994) Global design of analog cells using statistical optimization techniques. Analog Integr Circuits Signal Process 6(3):179–195
14. Castro-López R, Fernández F, Guerra-Vinuesa O (2006) Re-use based methodologies and tools in the design of analog and mixed-signal integrated circuits. Springer, Berlin
15. Stehr G, Pronath M, Schenkel F, Graeb H, Antreich K (2003) Initial sizing of analog integrated circuits by centering within topology-given implicit specification. In: Proceedings of the IEEE/ACM international conference on computer-aided design, pp 241–246
16. Balkir S, Dundar G, Alpaydin G (2004) Evolution based synthesis of analog integrated circuits and systems. In: Proceedings of NASA/DoD conference on evolvable hardware, pp 26–29
17. Takemura K, Koide T, Mattausch H, Tsuji T (2004) Analog-circuit-component optimization with genetic algorithm. In: Proceedings of the 47th midwest symposium on circuits and systems, vol 1, pp 489–492
18. Barros M, Neves G, Guilherme J, Horta N (2005) An evolutionary optimization kernel with adaptive parameters applied to analog circuit design. In: Proceedings of international symposium on signals, circuits and systems, vol 2, pp 545–548
19. Goh C, Li Y (2001) GA automated design and synthesis of analog circuits with practical constraints. In: Proceedings of the congress on evolutionary computation, vol 1, pp 170–177
20. Koza J, Bennett F III (1997) Automated synthesis of analog electrical circuits by means of genetic programming. IEEE Trans Evol Comput 1(2):109–128
21. Alpaydin G, Balkir S, Dundar G (2003) An evolutionary approach to automatic synthesis of high-performance analog integrated circuits. IEEE Trans Evol Comput 7(3):240–252

22. Kruiskamp W, Leenaerts D (1995) DARWIN: CMOS opamp synthesis by means of a genetic algorithm. In: Proceedings of the 32nd annual ACM/IEEE design automation conference, pp 433–438
23. Antao B, Gielen G, Rutenbar R (2002) Computer-aided design of analog integrated circuits and systems. Wiley, USA
24. Synopsys (2013) HSPICE homepage. http://www.synopsys.com/Tools/Verification/AMSVerification/CircuitSimulation/HSPICE/Pages/default.aspx
25. Storn R, Price K (1997) Differential evolution—a simple and efficient heuristic for global optimization over continuous spaces. J Glob Optim 11(4):341–359
26. Price K, Storn R, Lampinen J (2005) Differential evolution: a practical approach to global optimization. Springer, New York
27. Michalewicz Z (1996) Genetic algorithms+ data structures. Springer, New York
28. Deb K, Agrawal R (1995) Simulated binary crossover for continuous search space. Complex Syst 9(2):115–148
29. Poli R, Kennedy J, Blackwell T (2007) Particle swarm optimization. Swarm Intell 1(1):33–57
30. Rechenberg I (1994) Evolution strategy. Computational intelligence: imitating. Life 1:147–159
31. Hansen N (2006) The CMA evolution strategy: a comparing review. Towards a new evolutionary computation. Springer, pp 75–102
32. Gregory M, Bayraktar Z, Werner D (2011) Fast optimization of electromagnetic design problems using the covariance matrix adaptation evolutionary strategy. IEEE Trans Antennas Propag 59(4):1275–1285
33. Lagarias J, Reeds J, Wright M, Wright P (1998) Convergence properties of the Nelder-Mead simplex method in low dimensions. Siam J Optim 9:112–147
34. Chakraborty U (2008) Advances in differential evolution. Springer, Heidelberg
35. Liu B, Wang Y, Yu Z, Liu L, Li M, Wang Z, Lu J, Fernández F (2009) Analog circuit optimization system based on hybrid evolutionary algorithms. Integr VLSI J 42(2):137–148
36. Deb K (2000) An efficient constraint handling method for genetic algorithms. Comput Methods Appl Mech Eng 186(2):311–338
37. Medeiro F, Fernández F, Dominguez-Castro R, Rodriguez-Vazquez A (1994) A statistical optimization-based approach for automated sizing of analog cells. In: Proceedings of the IEEE/ACM international conference on computer-aided design, pp 594–597
38. Venkatraman S, Yen G (2005) A generic framework for constrained optimization using genetic algorithms. IEEE Trans Evol Comput 9(4):424–435
39. Deb K, Pratap A, Agarwal S, Meyarivan T (2002) A fast and elitist multiobjective genetic algorithm: NSGA-II. IEEE Trans Evol Comput 6(2):182–197
40. Zhang Q, Li H (2007) MOEA/D: a multiobjective evolutionary algorithm based on decomposition. IEEE Trans Evol Comput 11(6):712–731
41. Coello C, Lamont G, Veldhuizen D (2007) Evolutionary algorithms for solving multi-objective problems. Springer, New York
42. Li H, Zhang Q (2009) Multiobjective optimization problems with complicated Pareto sets, MOEA/D and NSGA-II. IEEE Trans Evol Comput 13(2):284–302
43. Zielinski K, Laur R (2006) Constrained single-objective optimization using differential evolution. In: Proceedings of IEEE congress on evolutionary computation, pp 223–230

Chapter 3
High-Performance Analog IC Sizing: Advanced Constraint Handling and Search Methods

This chapter concentrates on advanced constraint handling techniques and hybrid methods for high-performance analog circuit sizing. High-performance analog circuit sizing tools should have the following three properties: (1) ability to handle severe design specifications, (2) ability to obtain a highly optimized design, (3) ability to handle complex analog circuits. The fundamental techniques introduced in Chap. 2 are often not powerful enough for the above requirements. Therefore, advanced constrained optimization methods and hybrid methods are introduced, including state-of-the-art methods in the CI field, many of which have not been applied to the analog circuit sizing problem yet. Like Chap. 2, the method to make use of the CI methods to solve practical problems is emphasized by practical examples and analysis.

The remainder of the chapter is organized as follows. Section 3.1 reviews the challenges that are currently encountered for high-performance analog circuit sizing. Section 3.2 introduces the state-of-the-art constraint handling techniques. Section 3.3 introduces hybrid methods in the EC field. Section 3.4 provides a practical method for high-performance analog circuit sizing, called MSOEA. Section 3.5 summarizes this chapter.

3.1 Challenges in Analog Circuit Sizing

Although workable analog IC sizing methods can be developed using the techniques introduced in Chap. 2, the improvement of optimization algorithms for analog circuit sizing still remains an active research area because of the following reasons:

- More effective constraint handling techniques are greatly needed. The constraint handling problem is very important in analog circuit design, especially for high-performance (tough specifications) circuits. The drawbacks of the static penalty function-based method and the tournament selection-based method were described in Chap. 2, which will be further shown by practical examples in this chapter. To address this problem, advanced constrained optimization methods in the EC field

B. Liu et al., *Automated Design of Analog and High-frequency Circuits*,
Studies in Computational Intelligence 501, DOI: 10.1007/978-3-642-39162-0_3,
© Springer-Verlag Berlin Heidelberg 2014

need to be applied for the targeted problem. But many state-of-the-art constrained optimization methods have not been introduced into the EDA field yet.

- A more powerful search engine is needed to cope with more stringent specifications and to enhance the optimization ability. Genetic algorithms are good candidates due to their robustness and intrinsic parallelization capability. So it is the most popular evolutionary algorithm for analog IC sizing, but its search ability and convergence rate have been criticized [1]. It has also been proven that canonical GA cannot converge to the global optimum. GA with elitism converges to the global optimum theoretically, but this is not always the case in practice. In recent years, DE [2] and PSO [3, 4] are attracting much attention in the community of analog IC sizing and other similar areas (e.g., antenna design automation). EA aims at finding a satisfactory solution, instead of the exact global optimal solution. Hence, its local search ability needs enhancement. A possible solution is the use of memetic algorithms or hybrid methods. Their potentials in analog circuit design automation still need to be exploited.

3.2 Advanced Constrained Optimization Techniques

3.2.1 Overview of the Advanced Constraint Handling Techniques

The main challenge for constrained optimization compared to unconstrained optimization, is the trade-off of the objective function optimization and the minimization of the total constraint violation. They must be considered simultaneously in the evolutionary search, but the search direction of the two goals may be different in some period. In a population, some infeasible candidate solutions may be closer to the feasible global optimum than some current feasible solutions. To what extent that these promising infeasible solutions can be protected is determined by the penalty coefficients or the selection rules.

Let us use the tournament selection-based method [5] as an example. When no feasible solution is available in the current population, the evolution may follow a fast path to minimize the total violation of constraints without considering the objective function, causing some decision variable(s) swiftly enters one of the feasible intervals.[1] For simplicity, we assume the variable x_1 in the decision vector x has two feasible intervals: $[a_1, b_1]$ and $[a_2, b_2]$. For the current population, the total constraint violations when $x \in [a_1, b_1]$ is smaller than $x \in [a_2, b_2]$. After replacements in several iterations, x_1 of all of the individuals of the current population are in $[a_1, b_1]$. However, the optimal value of x_1 of the feasible global optimum may be within $[a_2, b_2]$. Unfortunately, x_1 can hardly converge to $[a_2, b_2]$ since the objective function

---

[1] There is an interval (or intervals) $[a', b']$ of a decision variable x_1 where values of x_1 of the feasible solutions are not located in them. The opposite is the feasible interval (or feasible intervals). Note that this is different from feasible region. A candidate solution cannot be guaranteed to be feasible even when all of the decision variables are in their feasible intervals.

has not been considered in the beginning and the candidates with $x_1 \in [a_2, b_2]$ has been replaced.

To address this problem, several advanced constrained optimization techniques are developed, including stochastic ranking-based methods, multi-objective optimization-based methods, Augmented Lagrangians and self-adaptive penalty function-based methods. They will be reviewed in this subsection.

- Stochastic ranking
  The stochastic ranking method [6] uses a probability factor p_f to determine whether the objective function value or the constraint violation value determines the rank of each individual. Therefore, the promising infeasible candidate solution has some probability to be selected. Experimental results show good performance of the proposed method. The main drawback is the difficulty to select p_f. In [7], an improved stochastic ranking method is proposed. Compared to the original stochastic ranking method using a constant p_f, the value of p_f decreases linearly from 0.475 in the initial generation to 0.025 in the final generation. An example using the idea of stochastic ranking in the tournament selection-based method is in [8]. At the beginning of the search process, more exploration of the search space is performed to find promising areas, while at the end of the search process, less infeasible solutions are allowed to keep the feasible solutions found and to discard the infeasible ones.
- Multi-objective optimization-based methods
  It is straightforward to see that the balance of the objective function optimization and the total constraint violation can be achieved by multi-objective optimization. Typical algorithms are [9] and [10]. [9] represents the methods treating each constraint as an objective. Good results can be achieved but the drawback is the high computational complexity, especially as the number of constraints increases. [10] uses a constraint satisfaction phase and an objective function optimization phase. In the constraint satisfaction phase, both the ranking of violation of constraints and the crowding distance are considered. In the objective optimization phase, the problem is treated as a bi-objective problem simultaneously optimizing the objective function and the total constraint violation.
- Augmented Lagrangians
  As has been discussed in Chap. 2, the static penalty function method suffers from the sensitivity of the penalty coefficients, but there is no general rule to assign them. To address this problem, a method is using Augmented Lagrangians, which dynamically determines the proper penalty coefficients. The method to cooperate augmented Lagrangians in EAs for constrained optimization is shown in [11, 12]. In this approach, the optimization problem is transformed into the minimization of an augmented Lagrangian. A competitive co-evolution method, composed of two evolutionary optimization processes running in parallel, is proposed to size analog circuits. The first evolutionary process tries to minimize the augmented Lagrangian, whereas the second one dynamically adapts the Lagrange multipliers (analogous to penalty coefficients) to optimum values. Similar method was used for high-performance analog circuit sizing [13]. Large improvements have been

shown, but the success is not guaranteed for high-performance analog circuit sizing problems with many severe constraints.

- Self-adaptive penalty function-based methods
  To determine the appropriate penalty coefficients, self-adaptive penalty function-based methods are developed [14, 15]. These methods aim at simultaneously considering the objective function optimization and total constraint violation minimization. It is straightforward that individuals with both low objective function value and low constraint violation value are better than individuals that have high objective function value or high constraint violation. It is desirable that more chances to remain are provided to those slightly infeasible solutions and having a low objective function value. Information extracted from the current population and previous populations is often used to assign the fitness. These methods often have good performance and are easily implemented. Moreover, the efficiency is also good. In the next subsection, a self-adaptive penalty function-based method will be further elaborated.

The above constraint handling techniques protect near feasible solutions with good objective function values in order to maintain the diversity. An alternative way is to produce more promising candidate solutions. [16] uses the standard tournament selection-based method to handle constraints, but the suitable learning strategy and parameter settings are gradually self-adapted according to the learning experience to generate more promising solutions. High performance was shown. [8] allows each solution to generate more than one offspring using different mutation operators to generate diverse promising candidates, and cooperates with the stochastic ranking method in the selection process to handle constraints.

3.2.2 A Self-Adaptive Penalty Function-Based Method

This subsection discusses a widely used normalization method and a self-adaptive penalty function-based method [14] for constrained optimization. This method is easy to implement and has high performance. No penalty function parameter needs to be defined by the user.

In the constraint satisfaction period, almost all constrained optimization methods minimize the total violation of constraints. In terms of implementation, normalization is often necessary, because different constraints may have different levels. For instance, the typical violation of a constraint $g_i(x)$ may be around 10^5, while that of another constraint $g_j(x)$ may be around 10. When they are sum up together without normalization, $g_i(x)$ will mainly affect the search direction, but the two constraints should be considered equally in the search. Therefore, the following normalization method is recommended, which is not limited to a certain kind of constraint handling method.

Recall that the constrained optimization problem we want to solve is as follows:

$$\begin{aligned}
\text{minimize}_x \quad & f(x) \\
\text{s.t.} \quad & g_i(x) \le 0, i = 1, \ldots, m \\
& x \in [x_L, x_H]
\end{aligned} \tag{3.1}$$

The violation of a constraint $g_i(x)$ can be defined as (3.2)

$$v_i(x) = \frac{c_i(x)}{cmax_i} \tag{3.2}$$

where $c_i(x) = max(0, g_i(x))$ and $cmax_i = max(c_i(x)), i = 1, \ldots, m$. In Eq. (2.1), there exist equality constraints $h(x)$, which usually refer to the KCL and KVL equations. They are inherently included in the simulation of analog ICs and do not appear explicitly. For other problems when equality constraints exist, $c_i(x) = max(0, |h_i(x) - \delta|)$. δ is the tolerance value for equality constraints. In such a way, each constraint can be considered equally in the evolutionary search process.

Now, we introduce the self-adaptive penalty function-based method [14]. Penalty function-based method uses fitness value considering the total constraint violation for ranking and comparison. The "self-adaptive penalty coefficients" in this method refers to the use of the information extracted from the current population and the current individual to select the penalty strategies. The fitness values of individuals for comparison or ranking are formulated by the sum of the distance value and the penalty value as shown by (3.3),

$$F(x) = d(x) + p(x) \tag{3.3}$$

where $d(x)$ is the distance value as formulated in (3.4) and $p(x)$ is the penalty value as formulated in (3.8).

$$d(x) = \begin{cases} v(x), & \text{if } r_f = 0 \\ \sqrt{f^*(x)^2 + v(x)^2}, & \text{otherwise} \end{cases} \tag{3.4}$$

In (3.4), there are $f^*(x)$, which is the normalized objective function and $v(x)$, which is the total constraint violation. The method to normalize the objective function is shown in (3.5).

$$f^*(x) = \frac{f(x) - f_{min}}{f_{max} - f_{min}} \tag{3.5}$$

The calculation of the total constraint violation is shown in (3.6).

$$v(x) = \frac{1}{m} \sum_{i=1}^{m} v_i(x) \tag{3.6}$$

(3.7) shows the calculation of the rate of feasible solutions in the current population.

$$r_f = \frac{\text{number of feasible individuals}}{\text{population size}} \tag{3.7}$$

From (3.4), it can be seen that:

(1) The distance value is equal to the constraint violation of the individuals when there is no feasible individual in the current population. This is the same as the tournament selection-based method in this period and is the best way of comparing infeasible individuals in the absence of feasible individuals.
(2) For feasible individuals, the distance value is equal to the normalized objective function value.
(3) For infeasible solution when there exist feasible solutions in the current population, the distance value is the square root of the normalized objective function value and the constraint violation. Therefore, if we compare two infeasible individuals based on their distance value, then the one that has both low objective function value and low constraint violation will be considered better-fit.
(4) For a feasible individual and an infeasible individual, either one can have smaller value. But if the two individuals have the same objective value, then the feasible individual will have smaller distance value. This protects promising infeasible solutions.

The penalty value composed of two factors is as follows.

$$p(x) = (1 - r_f)X(x) + r_f Y(x) \tag{3.8}$$

where

$$X(x) = \begin{cases} 0, & \text{if } r_f = 0 \\ v(x), & \text{otherwise} \end{cases} \tag{3.9}$$

$$Y(x) = \begin{cases} 0, & \text{if } x \text{ is a feasible individual} \\ f^*(x), & \text{otherwise} \end{cases} \tag{3.10}$$

From (3.8), it can be seen that: (1) when the number of feasible solutions in the population is small (but not zero), the first factor indicating the constraint violation will have more impact than the second penalty indicating the objective function optimization. (2) On the other hand if there are many feasible solutions in the population, the second factor has more effect than the first one.

For the combination of $d(x)$ and $p(x)$, it can be seen that:

(1) If there is no feasible individual in the current population, $d(x)$ will be equal to the constraint violation $v(x)$ and $p(x)$ will be zero.
(2) If there are feasible individuals in the population, then $d(x)$ will mainly determine which individuals are better fit.
(3) If two individuals have equal or very close distance value, then the penalty value $p(x)$ will determine which one is better.
(4) If there is no infeasible individual in the population, then individuals will be compared based on their objective function value alone.

3.3 Hybrid Methods

3.3.1 Overview of Hybrid Methods

For an evolutionary algorithm, its performance is determined by the trade-off between the exploration and exploitation throughout the run. Because of the "no free lunch" theorem, there is not a universal search mechanism or parameter setting that is advantageous over all kinds of problems (hyper-surfaces). In other words, if an EA outperforms another EA to some problems, then there must exist other problems showing the opposite performance. This illustrates the reason of why hybrid methods are needed. In the EC field, hybrid methods mainly focus on improving the solution quality and the convergence speed of a standard EA to many kinds of problems.

There are some widely used hybrid architectures [17]:

- Hybridization between an evolutionary algorithm and another evolutionary algorithm (for example: a genetic programming technique can be used to improve the performance of a genetic algorithm)
- Neural network assisted evolutionary algorithms
- Fuzzy logic assisted evolutionary algorithm
- Particle swarm optimization (PSO) assisted evolutionary algorithm
- Ant colony optimization (ACO) assisted evolutionary algorithm
- Bacterial foraging optimization assisted evolutionary algorithm
- Hybridization between evolutionary algorithm and other heuristics (such as local search, tabu search, simulated annealing, hill climbing, dynamic programming, greedy random adaptive search procedure, etc)

In these hybridizations, we mainly focus on the hybridization between multiple EAs and the hybridization of EA and other heuristics. In Sect. 3.5, an example will be provided to show the hybridization of real-coded GA and DE for high-performance analog IC sizing. In hybrid methods research, a very important topic is using local search in a standard EA, which is also called memetic algorithm (MA). MA has shown advantages on both the solution quality and the efficiency compared to standard EAs.

The general flow of an MA is as follows:

Step 1: Initialize the population of individuals by random generation.

Step 2: Evaluate the fitness of each individual in the initial population.

Step 3: Apply the evolution process until the termination condition is met:

    Step 3.1: Select the parent individuals for reproduction.

    Step 3.2: Generate offspring (child population) based on the parent individuals through crossover and mutation.

    Step 3.3: Evaluate the fitness of each individual in the child population.

    Step 3.4: Select the subset of individuals which undergo local refinement.

    Step 3.5: Perform local search (including evaluations) to the selected individuals.

Step 3.6: Update the subset of individuals when improvement is made.
Step 3.7: Update the population by replacing less fitted individuals by individuals with good fitness.

To clarify the above framework, we can take as an example the flow of a hybrid GA [18]. In this method, GA serves as the global exploration mechanism and the simulated annealing (SA) algorithm is used to perform local refinement of individual solution. In each iteration, GA operates on the current candidate solutions using selection, crossover, and mutation operators to produce offspring. Then, each of the offspring is sent to SA for further improvement. A single insertion neighborhood scheme is used. This process continues until all the offspring generated by GA are exhausted. At last, the best solutions of population size obtained from SA are the population for the next GA iteration.

3.3.2 Popular Hybridization and Memetic Algorithm for Numerical Optimization

A widely applied hybrid method is the hybridization of EA with simulated annealing (SA), such as cooperating GA and DE with SA [19]. SA is a probabilistic method proposed in Kirkpatrick, Gelett and Vecchi [20, 21] for finding the global minimum of a cost function that may possess several local minima. It works by emulating the physical process whereby a solid is slowly cooled so that when eventually its structure is "frozen", this happens at a minimum energy configuration.

The basic elements of simulated annealing (SA) are the following:

1. A finite set S.
2. A real-valued cost function J defined on S. Let $S^* \subset S$ be the set of global minima of the function J, assumed to be a proper subset of S.
3. For each $i \in S$, a set $S(i) \subset S - i$, called the set of neighbors of i.
4. For every $i \in S$, a collection of positive coefficients q_{ij}, $j \in S(i)$, such that $\sum_{j \in S(i)} q_{ij} = 1$. It is assumed that $j \in S(i)$ if and only if $i \in S(j)$.
5. A nonincreasing function $T(t)$: $N^+ \longrightarrow (0, \infty)$, called the cooling schedule. Here N^+ is the set of positive integers, and $T(t)$ is called the temperature at time t.
6. An initial state $x(0) \in S$.
7. Termination condition: such as a maximum number of iterations.

Given the above elements, the SA algorithm consists of a discrete-time inhomogeneous Markov chain $x(t)$, whose evolution we now describe. If the current state $x(t)$ is equal to i, choose a neighbor j of i at random; the probability that any particular $j \in S(i)$ is selected is equal to q_{ij}. Once j is chosen, the pseudocode to determine the next state $x(t + 1)$ is as follows:

If $J(j) \leq J(i)$,
$\qquad x(t + 1) = j$.

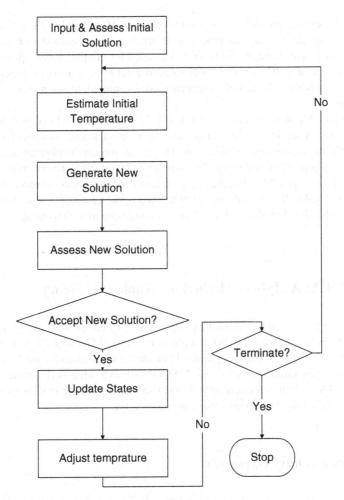

Fig. 3.1 Diagram of the simulated annealing algorithm

Elseif $J(j) > J(i)$
$\qquad x(t+1) = j$ with probability $Pr[x(t+1) = j | x(t) = i]$
End
Formally,

$$Pr[x(t+1) = j | x(t) = i] = q_{ij} exp\left[-\frac{1}{T(t)} max\{0, J(j) - J(i)\}\right]$$

If $\quad j \neq i \quad$ and $\quad j \notin S(i)$,   (3.11)
$\qquad Pr[x(t+1) = j | x(t) = i] = 0$

The flow diagram of the simulated annealing algorithm is shown in Fig. 3.1.

One of the critical problems in MA is the selection of the local exploitation mechanism. MA has shown large advantages on combinatorial optimization problems and many local search mechanisms have been used for combinatorial optimization. Nevertheless, analog IC sizing belongs to numerical optimization problem. In the following, an effective local refinement method for numerical optimization will be introduced.

The Nelder-Mead (NM) simplex method [22] is an effective local search mechanism in MA. The NM simplex algorithm is a direct search method for multidimensional unconstrained optimization. This method needs neither numerical nor analytical gradients. The NM method is easily understood and is described in Fig. 3.2. The hybridization of NM with EAs (e.g., DE and PSO) has received much attention [23, 24]. An analog IC yield optimization method using a memetic algorithm by DE and NM is also developed and show clear advantages compared to using a standard DE [25].

3.4 MSOEA: A Hybrid Method for Analog IC Sizing

Memetic single-objective evolutionary algorithm for high-performance analog IC sizing (MSOEA) [26] uses a hybridization of real-coded GA and DE as the search mechanism. A constraint handling method borrowing some ideas from SA is used to protect promising infeasible solutions. MSOEA achieves high performance for the targeted problem. In this section, MSOEA is used as an example to illustrate hybrid methods and advanced constraint handling techniques.

3.4.1 Evolutionary Operators

The DE algorithm [2] was introduced in Chap. 2. Compared with GA and PSO, DE has some attractive characteristics:

(1) The local search ability of DE is better than GA and some widely used evolutionary algorithms (EAs) [27];
(2) By using a one-to-one competition scheme to greedily select new candidates, knowledge of good solutions is retained in the current population, so elitism is ensured and results in a fast convergence character;
(3) All the operators work with real numbers, avoiding complicated generic search operators, so DE is efficient.

The drawback is that the diversity of the population decreases considerably in the evolution process when the population size is not large, so it may sometimes get premature solutions. However, a large population size will cause slow convergence.

The evolutionary operators of GA [28] have a large difference compared to DE. There are many different types of crossover, mutation and selection operators that

Input : $\gamma > 1$ (expansion factor), $\beta \in (0,1)$
                (contraction factor) and a tolerance ϵ

Output: Best solution found

repeat

Find x_h (worst point), x_l (best point), and x_g (the second worst point);

Compute the centroid: $x_c \leftarrow \frac{1}{n} \sum\limits_{i=1, i \neq h}^{n+1} x_i$;

Compute the reflected point: $x_r \leftarrow 2x_c - x_h$;

$x_{new} \leftarrow x_r$;

if $f(x_r) < f(x_l)$ **then**
  Make expansion: $x_{new} \leftarrow (1 + \gamma)x_c - x_h$;
else
  if $f(x_r) \geq f(x_h)$ **then**
    Make contraction: $x_{new} \leftarrow (1 - \beta)x_c + \beta x_h$;
  else
    if $f(x_g) < f(x_r) < f(x_h)$ **then**
      Make contraction:
      $x_{new} \leftarrow (1 + \beta)x_c - x_h$;
    end
  end
end

Compute $f(x_{new})$;

$x_h \leftarrow x_{new}$;

Compute $Q \leftarrow \left[\sum\limits_{i=1}^{n+1} \frac{(f(x_i) - f(x_c))^2}{n+1} \right]^{\frac{1}{2}}$;

until *Meeting criterion for termination: $Q < \epsilon$* ;

Fig. 3.2 The Nelder-Mead simplex method for local minimization

can construct different types of GAs. Binary GA [29] is a common EA in analog IC design optimization tools. However, binary GA suffers from Hamming cliffs in many cases, which affects the quality of the solutions [28]. On the other hand, some newly presented real-coded GA operators are shown to be effective, such as simulated binary crossover (SBX) [30], polynomial mutation and Gaussian mutation. In MSOEA, these operators will be used to find new candidates or to increase the population diversity.

Mutation Operator

MSOEA relies on the DE mutation operator [2] to explore the search space. The i-th individual in the d-dimensional search space at iteration t can be represented as

$$X_i(t) = [x_{i,1}, x_{i,2}, \ldots, x_{i,d}], i = 1, 2, \ldots, NP \tag{3.12}$$

where NP denotes the size of the population.

For each target individual i, according to the mutation operator, a mutant vector

$$V_i(t+1) = [v_{i,1}, v_{i,2}, \ldots, v_{i,d}], i = 1, 2, \ldots, NP \tag{3.13}$$

is generated by adding the weighted difference between a pair of individuals, randomly selected from the population at iteration t, to another individual, as described by the following equation:

$$V_i(t+1) = x_{r_0}(t) + F(x_{r_1}(t) - x_{r_2}(t)) \tag{3.14}$$

Details of the DE mutation operator are in Chap. 2. In this implementation, the base vector $X_{r_0}(t)$ is selected to be the best member of the current population, $X_{best}(t)$, so that the best information can be shared among individuals.

The second mutation operator is Gaussian mutation in real-coded GA [29]. For a parent x with standard deviation σ, a child is generated as follows:

$$\begin{aligned} \sigma' &= \sigma exp(N(0, \sigma)) \\ x' &= x + N(0, \sigma') \end{aligned} \tag{3.15}$$

where $N(0, \sigma)$ is a vector of independent random Gaussian numbers with zero mean and standard deviation σ. The cooperation of these two operators will be described later on.

Crossover operator There are two kinds of crossover operators in MSOEA. The first one is DE crossover [2]. For each target individual, a trial vector $U_i(t+1) = [u_{i,1}(t+1), u_{i,2}(t+1), \ldots, u_{i,d}(t+1)]$ is generated as follows:

$$u_{i,j}(t+1) = \begin{cases} v_{i,j}(t+1), & \text{if } (rand(i, j) \le CR)|j = randn(i) \\ x_{i,j}(t), & \text{otherwise} \end{cases} \tag{3.16}$$

More details of the DE crossover operator are in Chap. 2.

The second crossover operator is the simulated binary crossover (SBX) operator in real-coded GA. SBX has been introduced in Chap. 2. To make this section self-contained, the procedure of generating two children c_1 and c_2 from two parent solutions p_1 and p_2 using SBX is shown as follows:

1. Create a random number u between 0 and 1.
2. Calculate coefficient β:

$$\beta = \begin{cases} (2u)^{1/(\eta+1)} & \text{if } u \le 0.5 \\ \left(\frac{1}{2(1-u)}\right)^{1/(\eta+1)} & \text{, otherwise} \end{cases} \tag{3.17}$$

3. The children solutions are

$$\begin{aligned} c_1 &= 0.5[(1-\beta)p_1 + (1+\beta)p_2] \\ c_2 &= 0.5[(1+\beta)p_1 + (1-\beta)p_2] \end{aligned} \tag{3.18}$$

A small distribution index (η) would generate children solutions far away from the parent solutions, while a large η restricts children solutions to be near the parent solutions. The cooperation of these two operators will be described later on.

3.4.2 Constraint Handling Method

The constraint handling method in MSOEA is based on the tournament selection-based method [5]. Recall the three selection rules in Chap. 2 are:

(1) Given two feasible solutions, select the one with the better objective function value;
(2) Given two infeasible solutions, select the solution with the smaller constraint violation;
(3) If one solution is feasible and the other is not, select the feasible solution.

As has been said, the drawback is that some promising infeasible candidates may be lost in the search process, causing the final solution being trapped in a feasible region without the global optimum. In order to protect promising infeasible candidate solutions, and in combination with the one-to-one selection in DE, the above rule is modified by following a methodology inspired on acceptance with a probability in SA.

For a candidate x and its child x', suppose that one is feasible and the other is not. If the violation of the constraints of the infeasible solution is less than a predetermined bound Δ and it has a better objective function value, then it may be accepted according to a probabilistic criterion. This is analogous to a simulated annealing procedure: if the child is worse than the parent, SA does not necessarily reject it, but it may be accepted with a random finite probability. The use of this mechanism in MSOEA will be described in the next subsection.

3.4.3 Scaling Up of MSOEA

The flow diagram of the MSOEA algorithm is shown in Fig. 3.3. The first step is the population initialization. MSOEA is a real-coded EA, so it initializes the population as follows:

Fig. 3.3 Flow diagram of
MSOEA

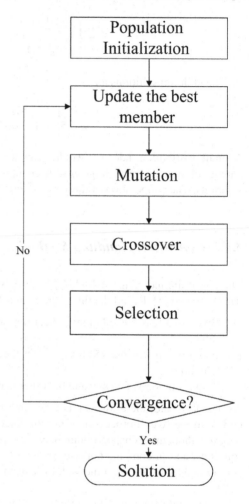

$$P = XV_{min} + rand(NP, d) \times (X_H - X_L) \qquad (3.19)$$

where P is the population; NP is the number of individuals in the population; d is
the number of decision variables; $rand(NP, d)$ is a matrix of uniformly distributed
random numbers between 0 and 1; and X_L and X_H are the vectors of minimum and
maximum values of the decision variables, respectively.

Then, the best member of the population is selected. Considering unconstrained
minimization problems, the best member is simply the point with minimum objective
function value in the current population. But for constrained optimization problems,
both the constraint satisfaction and objective minimization have to be considered.
The best member is, therefore, selected according to the rules in the tournament
selection-based method. Nevertheless, if a finite constraint violation Δ is permitted,

the above selection rule may provide an infeasible point as final solution. To solve this problem, the current best solution and the real best solution are defined. If there is no feasible solution, the one with the smallest constraints violation is the current best solution and the real best solution. If there are feasible solutions, the feasible solution with the smallest objective function value is the real best solution, and the solution with the smallest objective function value with a constraints violation below Δ is the current best solution. The current best solution is used in the evolution process, and the real best solution is stored and provided as final output. Sometimes, the two solutions may be the same.

Then, the mutation operator is applied, which uses the best member of the population, $X_{best}(t)$, as the base vector. At the beginning of the optimization process, the evolution mainly focuses on constraint satisfaction. $X_{best}(t)$ may have a relatively small constraint violation at this stage. By using $X_{best}(t)$ in the mutation operator, the generated individuals can share the good information on the small constraint violation according to the schema theorem [28]. At the end of the constrained optimization process, the algorithm focuses on objective minimization. $X_{best}(t)$ in this stage can provide good information on the optimized objective function value. Through experiments, it has been found that for practical analog circuit sizing problems, the search efficiency and effectiveness can be enhanced greatly by using $X_{best}(t)$ as the base vector for mutation (DE/best/1 mutation).

The second operator is the Gaussian mutation. However, unlike real-coded GA and the evolution strategy (ES), in the MSOEA algorithm, the purpose of Gaussian mutation is to increase the diversity of the population and to perform local tuning. Therefore, (3.20) is used to generate a certain number of candidates for an individual x in the population as follows:

$$x' = x + N(0, 0.01x) \tag{3.20}$$

For any of the candidates, if both the constraints violation and the objective function value of the new candidate are better than x, x will be updated.

Notice that the usefulness of the two mutation operators is remarkably different. The DE mutation operator dominates the exploration and exploitation of the solution space, and it is used at every iteration, whereas the Gaussian mutation operator is a supplement for increasing population diversity and local tuning, so it is only used after the mean violation of constraints is less than Δ, and used once every GM iterations. GM values between 10 and 20 are appropriate to achieve the desired goal.

Next, the DE crossover operator is applied. In the DE algorithm, the newly generated vector is only selected from the original vector and the mutant vector. So the SBX crossover operator from real-coded GA is added to jump out of premature convergence by introducing another exploitation mechanism, and is executed every XC iterations. XC values around 10 are appropriate to achieve the desired diversity increase.

Finally, selection takes place. In MSOEA, however, the tournament selection procedure to handle constraints has been modified with a probabilistic acceptance method inspired by SA. If a child has a better value of the objective function and lies close to the feasible region (with a constraints violation below Δ), then this solution may be good for final optimization, even though it is not actually feasible. Therefore, it can be accepted with a certain probability. Here, the same principle but a different style of the updating rule of probabilistic acceptance like in SA is used. An infeasible solution with a constraint violation below Δ is accepted if the following condition is satisfied:

$$exp(-(obj_p - obj_s)/|obj_p|) \leq OT + rand(i) \times (1 - OT) \qquad (3.21)$$

where obj_p and obj_s correspond to the parent's and child's objective value, and $rand(i)$ is a uniformly distributed random number between 0 and 1. Parameter OT controls what improvement of the objective function is automatically selected. For instance, $OT = 0.9$ implies that infeasible solutions with constraints violation below Δ are automatically selected if the objective function improvement is larger than $ln(OT) = 10.53\%$ and randomly accepted if the objective improvement is below 10.53%.

If a stop criterion is met (e.g., a convergence criterion is satisfied, or a maximum number of iterations is reached), then the algorithm outputs $X_{best}(t)$ and its objective function value; otherwise a new generation is created.

This optimization method exhibits the following features:

- The mutation and one-to-one selection mechanism of differential evolution are basic operators in MSOEA. Therefore, the local search ability, elitism and efficiency of this algorithm are directly inherited.
- The incorporation of the SBX crossover and Gaussian mutation operators promote local search while maintaining a high diversity for global search. Moreover, the formulation of the standard deviation in the Gaussian mutation operator implements a self-adaptive mechanism.
- The proposed selection mechanism for the base vector of the mutation operator in a constrained optimization problem contributes to increasing the search efficiency and effectiveness considerably.
- No penalty coefficients are needed to handle design constraints.
- The selection rules are revised to preserve potentially good infeasible solutions close to the feasible region. This contributes to improving the solution quality.

3.4.4 Experimental Results of MSOEA

3.4.4.1 Example 1: Two-Stage Telescopic Cascode Amplifier

A two-stage telescopic cascode amplifier, shown in Fig. 3.4, is used as an example. The load capacitance is 1.6 pF, and the technology is a 0.25 μm CMOS process with 2.5 V power supply. Transistor lengths were allowed to vary between the minimum value allowed by the technological process, 0.25 μm, up to 10 μm. Transistor widths

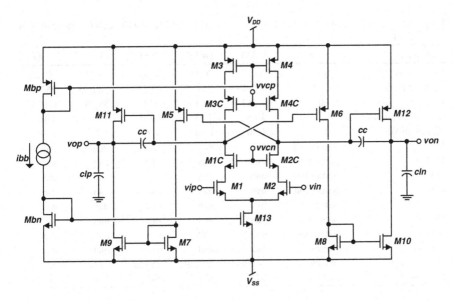

Fig. 3.4 CMOS two-stage telescopic cascode amplifier

were changed between the minimum technology value, 0.5 μm, up to 1000 μm. The bias current was allowed to vary between 0.5 μA and 10 mA and the bias voltages from 50 mV to 1.1 V. The capacitor values could change from 100 fF to 50 pF. Appropriate matching relations were imposed: $M1 \equiv M2$, $M1c \equiv M2c$, $M3 \equiv M4 \equiv Mbp$, $M3c \equiv M4c$, $M13 \equiv 2 \times Mbn$, $M6 \equiv M5$, $M8 \equiv M7$, $M10 \equiv M9$, $M12 \equiv M11$. The minimization goal is the area occupation whereas the specifications of other performances are quite high.

Four reference methods are used, including (1) the genetic algorithm, combined with penalty functions to handle constraints (GAPF); (2) the differential evolution algorithm as search engine and the same penalty-based method to handle constraints (DEPF); (3) the competitive co-evolutionary differential evolution algorithm CODE [13]; and (4) an algorithm that combines the selection-based method for constrained optimization proposed by [5] and differential evolution (denoted as SBDE) [31]. Table 3.1 shows the results achieved by the different algorithms. It can be seen that all the constraints are successfully met only by MSOEA.

To test the robustness of the different algorithms, 20 runs of each of them were executed. For the GAPF and DEPF algorithms, all constraints were normalized with respect to the specified value and equal weights were assigned to all of them. Table 3.2 shows the results for the different methods. The first two columns show the number of feasible and infeasible solutions found in the 20 runs. It can be seen that the GAPF and DEPF algorithms were never able to find a feasible solution. The constraints are very hard to meet and all algorithms find difficulties. However, whereas most times the SBDE and CODE algorithms were not able to find a feasible solution (only 2 and 3 times out of 20), MSOEA was able to find it most times (17 out of 20). The

Table 3.1 Design specifications and results of GAPF, DEPF, SBDE, CODE and MSOEA (example 1)

Specifications	Constraints	GAPF	DEPF	SBDE	CODE	MSOEA
DC gain (dB)	≥ 80	70.17	70.85	76.68	81.22	80.34
GBW (MHz)	≥ 250	250.17	249.98	250.33	280.30	252.85
Phase margin (°)	≥ 60	40.83	54.16	47.12	58.49	60.24
Output swing (V)	≥ 4	3.55	4.58	4.61	3.86	4.16
Power (mW)	≤ 4	6.20	6.34	3.99	4.76	3.78
M1	Saturation	met	met	met	met	met
M1C	Saturation	met	met	met	met	met
M3	Saturation	met	met	met	met	met
M3C	Saturation	met	met	met	met	met
M5	Saturation	met	met	met	met	met
M9	Saturation	met	met	met	met	met
M11	Saturation	met	met	met	met	met
M13	Saturation	met	met	met	met	met
Area (μm^2)	objective	14506	2642.9	2582.2	2217.7	3129.3
Time (s)	N.A.	572	551	546	590	566

Table 3.2 Statistical results on constraint satisfaction and design objective (example 1)

Item	Feasible	Infeasible	Best	Worst	Average
GAPF	0	20	N.A.	N.A.	N.A.
DEPF	0	20	N.A.	N.A.	N.A.
CODE	3	17	3643.9	4171.9	3887.9
SBDE	2	18	3205.8	7910.0	5557.9
MSOEA	17	3	2522.1	4212.0	3238.3

following three columns in Table 3.2 show the best, worst and average value of the area occupation of the feasible solutions.

The results of this example with extremely stringent specifications show that both the search engine and the constraint handling technique determine the quality of the solutions. Penalty-based methods exhibit large difficulties in finding a feasible solution. A possible reason for the inferior solutions of CODE is that the search engine is merely DE, although the Augmented Lagrangians is an advanced constraint handling method. MSOEA, on the other hand, can maintain its search ability by relying on both the DE operators and other memetic operators that palliate the potential shortcomings of DE. Moreover, the mechanism to preserve promising infeasible solutions contributes to the quality of the final MSOEA results.

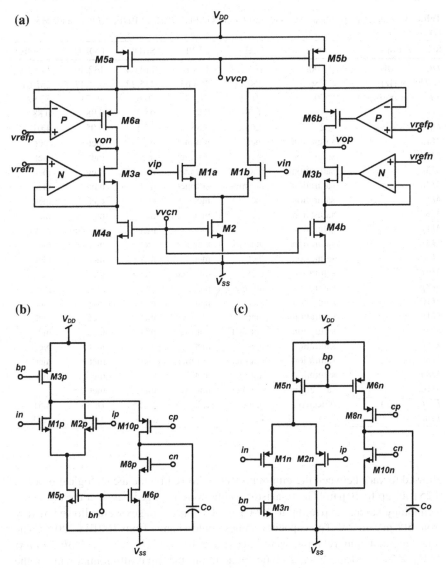

Fig. 3.5 a CMOS gain-boosted folded-cascode amplifier; **b** P amplifier in **a**; and **c** N amplifier in **a**

3.4.4.2 Example 2: Gain-Boosted Folded-Cascode Amplifier

The gain-boosted folded-cascode amplifier in Fig. 3.5 is optimized using MSOEA in this example. This circuit has 44 design parameters and 23 constraints (including those necessary to ensure the proper operating region for all transistors). A 0.25 μm CMOS technology with 2.5 V power supply is used. The transistor lengths were

Table 3.3 Design specifications and results of GAPF, DEPF, SBDE, CODE and MSOEA (example 2)

Specifications	Constraints	GAPF	DEPF	SBDE	CODE	MSOEA
DC gain (dB)	≥ 100	105.55	123.14	114.74	132.40	123.10
GBW (MHz)	≥ 250	250.14	253.63	335.22	254.47	255.61
Phase margin (°)	≥ 70	70.11	71.30	70.63	79.71	72.82
gm	< 1	0.99	0.95	0.99	0.98	0.88
Output swing (V)	≥ 2.5	2.64	3.85	3.72	3.05	4.01
*M*1*a*	saturation	met	not met	met	met	met
*M*2	saturation	met	not met	met	met	met
*M*3*a*	saturation	not met	not met	met	met	met
*M*4*a*	saturation	not met	not met	met	met	met
*M*5*a*	saturation	met	met	met	met	met
*M*6*a*	saturation	met	met	met	met	met
*M*1*p*	saturation	met	met	met	not met	met
*M*3*p*	saturation	not met	met	met	met	met
*M*5*p*	saturation	met	met	met	met	met
*M*6*p*	saturation	met	not met	met	met	met
*M*8*p*	saturation	not met	met	met	met	met
*M*10*p*	saturation	met	met	met	met	met
*M*1*n*	saturation	met	not met	met	met	met
*M*3*n*	saturation	not met	not met	met	met	met
*M*5*n*	saturation	met	met	met	met	met
*M*6*n*	saturation	not met	not met	met	met	met
*M*8*n*	saturation	not met	met	met	met	met
*M*10*n*	saturation	not met	not met	met	met	met
Power (mW)	objective	6.36	1.23	11.55	18.33	6.51
Time (s)	N.A.	1299	1165	1018	3703	1028

allowed to vary between the minimum value allowed by the technological process, 0.25 μm, up to 10 μm. The transistor widths were changed between the minimum technology value, 0.5 μm, up to 1000 μm. The bias voltages were allowed to vary from 500 mV to 2 V. The capacitor values could change from 100 fF to 20 pF. The following matching relations were imposed: $M1a \equiv M1b$, $M3a \equiv M3b$, $M4a \equiv M4b$, $M5a \equiv M5b$, $M6a \equiv M6b$. In addition, the input differential pairs in the boosting amplifiers are matched and obviously, the two boosting amplifiers of each type are identical. The population size was set to 80 for all algorithms.

The design specifications and typical results of GAPF, DEPF, CODE, SBDE and MSOEA are shown in Table 3.3. It can be seen that GAPF and DEPF do not meet the performance specifications and that several transistors are out of the saturation region. The CODE algorithm was able to satisfy the constraints except for a small violation of the saturation constraint for a pair of transistors (e.g., M1p). SBDE, like MSOEA, satisfied all the constraints, but the power dissipation is twice that achieved by MSOEA.

The statistics of some operators used in MSOEA are interesting to look at. In a typical runs, the selection operator with probabilistic acceptance of infeasible points is active 1088 times in a typical run of this example. This means that many solutions that have a very good objective function and are very close to the feasible region have been protected and potentially kept in the population. On the other hand, the Gaussian mutation operator generates 826 better individuals (representing about 3.8 % of the total number of mutations), improving the local tuning and increasing the diversity of the population.

3.5 Summary

This chapter has introduced advanced CI techniques for constrained optimization, which are especially useful for global optimization problems with severe constraints, e.g., high-performance analog IC sizing. Advanced constraint handling methods and hybrid methods have been introduced in this chapter for the targeted problem. Advanced constraint handling method can be classified into stochastic ranking method, multi-objective optimization-based method, Augmented Lagrangians and self-adaptive penalty function-based methods. An effective and easy to implement self-adaptive penalty function-based method has been elaborated. For hybrid methods, the hybridization of evolutionary operators in different EAs and the hybridization of evolutionary operators and local search methods (memetic algorithms) have been introduced. The NM simplex method has been recommended as the local search mechanism for numerical optimization. Note that advanced constrained handling methods and hybrid methods often cooperate and help with each other. An example method, MSOEA, has been provided. There are many possibilities to appropriately combine advanced constraint handling methods and hybrid methods to address the targeted problem, which have good potential to be explored.

References

1. Rudolph G (1994) Convergence analysis of canonical genetic algorithms. IEEE Trans Neural Networks 5(1):96–101
2. Price K, Storn R, Lampinen J (2005) Differential evolution: a practical approach to global optimization. Springer-Verlag, New York
3. Kennedy J (2006) Swarm intelligence. In Handbook of nature-inspired and innovative computing, Springer, Heidelberg, pp 187–219
4. Fakhfakh M, Cooren Y, Sallem A, Loulou M, Siarry P (2010) Analog circuit design optimization through the particle swarm optimization technique. Analog Integr Circ Sig Process 63(1):71–82
5. Deb K (2000) An efficient constraint handling method for genetic algorithms. Comput Methods Appl Mech Eng 186(2):311–338
6. Runarsson T, Yao X (2000) Stochastic ranking for constrained evolutionary optimization. IEEE Trans Evol Comput 4(3):284–294

7. Mallipeddi R, Suganthan P (2010) Ensemble of constraint handling techniques. IEEE Trans Evol Comput 14(4):561–579
8. Mezura-Montes E, Velázquez-Reyes J, Coello C (2006) Modified differential evolution for constrained optimization. In Proceedings of IEEE congress on evolutionary computation, pp 25–32
9. Wang Y, Cai Z, Guo G, Zhou Y (2007) Multiobjective optimization and hybrid evolutionary algorithm to solve constrained optimization problems. IEEE Trans Syst Man Cybern Part B: Cybern 37(3):560–575
10. Venkatraman S, Yen G (2005) A generic framework for constrained optimization using genetic algorithms. IEEE Trans Evol Comput 9(4):424–435
11. Tahk M, Sun B (2000) Coevolutionary augmented lagrangian methods for constrained optimization. IEEE Trans Evol Comput 4(2):114–124
12. Barbosa H (1999) A coevolutionary genetic algorithm for constrained optimization. In Proceedings of the congress on evolutionary computation, pp 1605–1611
13. Liu B, Wang Y, Yu Z, Liu L, Li M, Wang Z, Lu J, Fernández F (2009) Analog circuit optimization system based on hybrid evolutionary algorithms. Integration, VLSI J 42(2):137–148.
14. Tessema B, Yen G (2006) A self adaptive penalty function based algorithm for constrained optimization. In Proceedings of IEEE congress on evolutionary computation, pp 246–253
15. Farmani R, Wright J (2003) Self-adaptive fitness formulation for constrained optimization. IEEE Trans Evol Comput 7(5):445–455
16. Huang V, Qin A, Suganthan P (2006) Self-adaptive differential evolution algorithm for constrained real-parameter optimization. In Proceedings of IEEE congress on evolutionary computation, pp 17–24
17. Grosan C, Abraham A, Ishibuchi H (2007) Hybrid evolutionary algorithms. Springer, Verlag
18. Ganesh K, Punniyamoorthy M (2005) Optimization of continuous-time production planning using hybrid genetic algorithms-simulated annealing. Int J Adv Manuf Technol 26(1):148–154
19. Hu Z, Su Q, Xiong S, Hu F (2008) Self-adaptive hybrid differential evolution with simulated annealing algorithm for numerical optimization. In Proceedings of IEEE world congress on computational intelligence, pp 1189–1194
20. Bertsimas D, Tsitsiklis J (1993) Simulated annealing. Stat Sci 8(1):10–15
21. Kirkpatrick S, Gelatt C Jr (1983) Optimization by simulated annealing. Science 220(4598):671–680
22. Lagarias J, Reeds J, Wright M, Wright P (1998) Convergence properties of the Nelder-Mead simplex method in low dimensions. SIAM J Optim 9:112–147
23. Zahara E, Kao Y (2009) Hybrid Nelder-Mead simplex search and particle swarm optimization for constrained engineering design problems. Expert Syst Appl 36(2):3880–3886
24. Menchaca-Mendez A, Coello C (2009) A new proposal to hybridize the Nelder-Mead method to a differential evolution algorithm for constrained optimization. In Proceedings of IEEE congress on evolutionary computation, pp 2598–2605
25. Liu B, Fernández F, Gielen G (2010) An accurate and efficient yield optimization method for analog circuits based on computing budget allocation and memetic search technique. In Proceedings of the conference on design, automation and test in Europe, pp 1106–1111
26. Liu B, Fernández F, Gielen G, Castro-López R, Roca E (2009) A memetic approach to the automatic design of high-performance analog integrated circuits. ACM Transactions on Design Automation of Electronic Systems (TODAES) 14(3):1–24
27. Storn R, Price K (1997) Differential evolution–a simple and efficient heuristic for global optimization over continuous spaces. J Glob Optim 11(4):341–359
28. Michalewicz Z (1996) Genetic algorithms+ data structures. Springer, New York.
29. Gen M, Cheng R (2000) Genetic algorithms and engineering optimization. Wiley-interscience, New York
30. Deb K, Agrawal R (1994) Simulated binary crossover for continuous search space. Complex Syst 1(9):115–148
31. Zielinski K, Laur R (2006) Constrained single-objective optimization using differential evolution. In Proceedings of IEEE congress on evolutionary computation, pp 223–230

Chapter 4
Analog Circuit Sizing with Fuzzy Specifications: Addressing Soft Constraints

4.1 Introduction

In Chaps. 2 and 3, we have introduced analog circuit sizing methods with crisp or deterministic constraints. In this chapter, we introduce the solution method to analog IC sizing problems with fuzzy constraints. A deterministic or hard constraint means that any violation of the constraint is not acceptable whereas a fuzzy constraint is a kind of soft constraint, which means a little violation of the constraint may also be acceptable. For instance, for the constraint DC gain ≥ 3 dB, a 2.9 dB DC gain may be acceptable if it is considered as a soft constraint. In this chapter, two fuzzy constraint handling methods are introduced. Single and multi-objective analog IC sizing methods with soft constraints are then presented, respectively. Combining the fuzzy constraint handling method for single-objective optimization and the differential evolution (DE) algorithm as the search engine, the fuzzy selection-based differential evolution (FSBDE) method is presented for single-objective fuzzy analog IC sizing. Combining the fuzzy constraint handling method for multi-objective optimization and the non-dominated sorting genetic algorithm II (NSGA-II) as the search engine, the multi-objective fuzzy selection-based analog circuit sizing method (MOFSS) is presented.

The remainder of this chapter is organized as follows. Section 4.2 introduces the motivation for using fuzzy specifications. Section 4.3 discusses how to construct fuzzy numbers for a certain application. Section 4.4 introduces a fuzzy selection-based constraint handling method for single-objective evolutionary algorithm-based optimization. Its combination with DE to construct the FSBDE method is introduced in Sect. 4.5. The MOFSS method and the revised fuzzy selection rules for multi-objective analog circuit sizing, is introduced in Sect. 4.6. Section 4.7 summarizes this chapter.

B. Liu et al., *Automated Design of Analog and High-frequency Circuits*,
Studies in Computational Intelligence 501, DOI: 10.1007/978-3-642-39162-0_4,
© Springer-Verlag Berlin Heidelberg 2014

4.2 The Motivation of Analog Circuit Sizing with Imprecise Specifications

4.2.1 Why Imprecise Specifications Are Necessary

In both the CI and the analog circuit sizing fields, most available constraint handling methods are designed to handle crisp constraints. This is indeed necessary in many cases. Nevertheless, in some other cases (such as the investigation of a new circuit configuration), this is a too rigid approach to capture actual human intentions [1]. The designer is often interested in the neighborhood of the obtained solution; he /she would like to know which constraint can have an acceptable relaxation in order to obtain a better objective function or trade-off curves, rather than strictly conforming to the, sometimes arbitrarily, pre-defined specifications. For instance, it is possible that the designer would like to know if the power would be significantly improved when the phase margin of an amplifier is slightly lower than 60° (assuming that the specification is phase margin ≥60°). But such information cannot be provided by methodologies based on crisp constraints. In crisp or deterministic constraint-based optimization, the first requirement is to satisfy the constraints. For example, in the tournament selection-based method, feasible solutions are always better than infeasible solutions. Therefore, a candidate solution with a phase margin of 59° and 1 mW power consumption is regarded to be worse than a candidate solution with 60° and 1.5 mW power consumption. This is against the real intention of the designer in many occasions. Not only in analog IC sizing, this situation occurs in many other application areas. On the other hand, the soft relaxation of some performances is only useful when other interesting performances can be enhanced and the relaxation must be small enough. Simply relaxing the constraints makes little sense to this problem.

Soft constraints or specifications are applicable to both single- and multi-objective analog circuit sizing. For single-objective sizing, the reason is described in the last paragraph. Besides using soft constraints, another method to include human intentions into single-objective analog circuit sizing which needs to be mentioned is to set all performances as design objectives and apply a multi-objective optimization algorithm. By using multi-objective analog IC sizing [2–9], the trade-offs between the different objectives can be explored. Then, the decision on an appropriate balance of performances can be taken a posteriori. However, the major drawbacks of using multi-objective optimization to address this problem are that: (1) A lot of computational resources are wasted in getting solutions in regions of the performance space without any current interest for the designer if he or she just wants a single-objective sizing. (2) Available multi-objective evolutionary algorithms often cannot handle problems with more than five objectives very well. In contrast, if multi-objective optimization is used to catch imprecise human intentions, the number of objectives is often more than five. Hence, effective soft constraint handling methods in single-objective analog circuit sizing are greatly needed.

A question is that is soft constraint also necessary if the designer wishes a multi-objective sizing? The answer is yes. In multi-objective sizing, performances can either be set as objectives or constraints. An objective means that the designer is interested in its trade-offs with all other objectives in the whole performance space, which contains much information; while a constraint usually means that the performance needs to be "larger than" or "smaller than" a certain value, and no trade-off information is considered. Usually, only a part of the important performances are considered as objectives, and the others can be set as constraints. The reason is that the designer is often not interested in the trade-offs among all objectives. Moreover, there are typically other design decisions, such as restricting the transistors to operate in the saturation region, which are essentially defined as functional constraints. However, there are very few works which study the constraint handling technology in multi-objective analog circuit sizing.

In multi-objective sizing, it is possible that the designer would like to know if the power versus area trade-off curve would be significantly improved when the phase margin is slightly lower than 60° (assuming the specification is phase margin ≥60°). If we define the trade-off between the objectives in the whole performance space as Pareto front-based trade-off, the trade-off between the neighborhoods of the constraint boundaries can be defined as local trade-off. Currently, with crisp constraints, the only way to explore the local trade-offs is by repeatedly modifying the constraint boundaries. For example, the designer may first reduce the 60° phase margin specification to 58° and check the results. If satisfactory, he / she may further decrease the constraint boundary to 55°, which could be the lowest specification considered and compare the results; otherwise, 60° will be retained. However, more than one specification may be involved in the local trade-offs in a real design, and the combination of them may result in numerous tunings.

Consequently, a good soft constraint handling technology is very important in both single- and multi-objective analog IC sizing methods. This chapter will investigate both of them.

4.2.2 Review of Early Works

Fuzzy sets theory and fuzzy logic have appeared in the analog circuit synthesis field [1, 10–13], covering various topics, e.g., symbolic modeling, optimization, topology selection, etc. Fuzzy constraints-based analog circuit sizing methodologies are described in [1, 10].

The methods in [1, 10] take into account performance tolerances and allow varying degrees of acceptability of a particular solution. By using fuzzy numbers to represent constraints, imprecision and vagueness of the designer's intentions are supported in analog synthesis.

For fuzzy constraints and objectives, membership functions $\mu(x)$ are defined, reflecting the fulfillment of the fuzzy constraints $g(x)$ or the objective $f(x)$. The constrained optimization problem is transformed into the maximization of the weighted sum of the membership function values, which can be described by the following equation:

$$\mu(x) = \sum_{i=1}^{m} w_i \mu_{fi}(x) + \sum_{j=1}^{n} w_j \mu_{gj}(x) \tag{4.1}$$

where m is the number of objectives, n is the number of inequality constraints, $\mu_{fi}(x)$ and $\mu_{gj}(x)$ are their membership function values and w_i and w_j are the weights to prioritize objectives and constraints. The membership function for constraints (considering $g_j(x) \leq spec_j$) in the reported methods is defined as [1]:

$$\mu_{gj}(x) = \begin{cases} 1, & \text{if } g_j(x) \leq spec_j \\ exp\left(-\left(\frac{g_j(x)-spec_j}{p_j/2}\right)^2\right), & \text{otherwise} \end{cases} \tag{4.2}$$

where $p_j > 0$ is a tolerance interval, that can be controlled by the designer.

The advantage of these methods is that they allow tolerance intervals for the constraints, which solves the problems of using crisp constraints. However, some drawbacks limit their applicability. Firstly, a weighted-sum method is used. The result is very sensitive to the weights, but there is no systematic approach to choose the proper weights. In other words, it leaves uncertainty to the designer whether the chosen weights appropriately reflect his preferences [14]. Secondly, the fuzzy number construction itself, illustrated in Fig. 4.1, has limitations. Suppose that the specification of DC gain is 60 dB. First, as shown in Fig. 4.1a, a large p_j provides a large span of the fuzzy sets, which makes $\mu(x) > 0$ for values quite far from the designer's specification. This may cause the sizing to be "too soft". In contrast, in Fig. 4.1b, a small p_j reflects the designer's intentions well, but this will highly affect the evolution. Indeed, in the first few iterations, it is possible that no candidate reaches the $\mu(x) > 0$ region of any fuzzy number, so all the fitness values are 0 in Eq. (4.2), and the evolution will fail.[1] This aspect is further elaborated in Sect. 4.4 and is demonstrated in Sect. 4.5 by experiments.

4.3 Design of Fuzzy Numbers

This section introduces a fuzzy number construction method considering the handling of imprecise human intentions and the evolutionary search-based optimization.

The shortcoming of crisp constraints and the advantage of imprecise human intentions can be shown by the following example. Suppose that the specification on the phase margin is 60° and the objective is power minimization. Assume there are two candidate solutions. The first one has a 59° phase margin and a power consumption

---

[1] Actually, $\mu(x)$ is not exactly zero when using an exponential membership function, but there is a large region with $\mu(x) \simeq 0$ and there is no distinct improvement, as shown by the points with DC gain less than 40 dB in Fig. 4.1b. The fitness values of these points have little difference, and from these points it is difficult to reach regions with $\mu(x)$ distinctly larger than zero. Therefore, $\mu(x) = 0$ is used for this condition in the following.

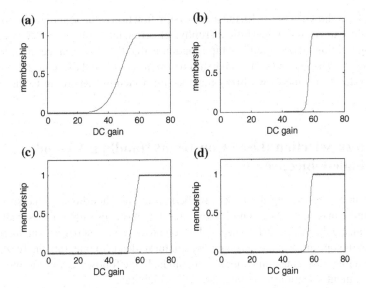

Fig. 4.1 Membership functions

of 1.5 mW, while the second one has 65° and 2.5 mW, respectively. According to the crisp constraint handling method, the first solution will be eliminated because it does not meet the strict phase margin specification of 60°. However, most designers will choose the first one in practice.

In this chapter, fuzzy numbers (also called "fuzzy sets") are integrated in the optimization process to address this human intention. A fuzzy number is a set of values, each having a membership function $\mu(x)$ that reflects the degree of acceptability of the required value. $\mu(x) = 1$ means that the value is fully acceptable, and $\mu(x) = 0$ means that the value is not acceptable at all. For more details about fuzzy numbers and fuzzy sets, please see Chap. 1.

The following two kinds of fuzzy numbers are shown to be useful [15, 16]. The first one is a linear fuzzy number, as shown in Fig. 4.1c. The membership increases linearly in the $\mu(x) \geq 0$ and $\mu(x) \leq 1$ area. The designer needs to specify the upper and lower bounds of the specifications. The advantage of a linear fuzzy number is that it is straightforward and easy to design, and can reflect the designer's intentions in many cases. The drawback is that its slope is a constant, while the designer may need a higher slope in some region and a lower slope in another region. Therefore, the fuzzy number from [1, 10] (Eq. 4.2) is also used, but two modifications are applied. Firstly, p_j is restricted to a certain interval, in order to restrict the fuzzy number to be within the neighborhood of the specifications. Secondly, upper and lower bounds also exist. Take the constraint of DC gain $\geq ga$ for example, for values below the lower bound, μ_x is set to 0.[2] Although the membership function is not continuous at

---

[2] For the specifications which can be presented by smaller than the upper bound, it can be transformed to the above style by simply adding a negative sign.

$\mu(x) = 0$, as shown in Fig. 4.1d, the membership function value near the lower bound is very close to zero, so it has little disruptive effect on the evolution procedure. The advantage of this kind of membership function is that the shape can be modified by changing p_j. The above two membership functions are typical examples; designers can also decide on their own fuzzy membership functions according to their own needs.

4.4 Fuzzy Selection-Based Constraint Handling Methods (Single-Objective)

As said above, restricting the fuzzy numbers to the neighborhood of the specified constraints can reflect the designer's intentions, but may cause all the fitness values to be zero during the evolution because of the weighted-sum method used, and therefore may ruin the optimization. Consequently, a natural idea is to integrate the fuzzy sets with available crisp constraint handling methods. In the following, an example using the tournament selection-based method [17] is elaborated.

For a set of constraints $\{g_1(x) \geq 0, g_2(x) \geq 0, \ldots, g_n(x) \geq 0\}$, the constraint violation is calculated as

$$v_c = \sum_{i=1}^{n} |min(g_i(x), 0)| \tag{4.3}$$

When using the selection-based constraint handling method, two candidates with $\sum \mu(x) = 0$ can still be compared by their violation of the constraints to direct the selection procedure in the evolution process.

In order to derive the selection rules, the designer's intentions are summarized as follows:

(1) The designer may allow a certain extent of tolerance for the specifications, but only in the neighborhood of the specifications.
(2) The designer may only accept constraint degradation when the objective function value can effectively be improved. If there is no improvement to the objective function, the relaxation of the constraints is of little sense, because the new solution is inferior to the solution that did not allow tolerance on the objective and constraints.
(3) Solutions that have many performances with $\mu(x) > 0$ but close to $\mu(x) = 0$ may appear in the sizing result. Although being in the allowed region, such solutions with obviously many degradations can hardly be acceptable for the designer.
(4) Although the designer can allow a tolerance, he may also wish that the performances are forced to meet the specifications if the objective function value is not harmed.

To address the third rule, besides the allowed region of the fuzzy number (the lower and upper bounds), a λ-cut set [18] is used. A λ-cut set is a clear set A_λ from the fuzzy

Table 4.1 Fuzzy identifications of the performances

Level of performances	Judgment
Fully satisfactory	$\mu(x) = 1$
Comparatively satisfactory	$\mu(x) \geq \lambda$ and $\mu(x) < 1$
Allowed but not satisfactory	$\mu(x) > 0$ and $\mu(x) < \lambda$
Not acceptable	$\mu(x) = 0$

set, which can be expressed as $A_\lambda = \{x | \mu(x) \geq \lambda\}$. When the obtained performance of the candidate has a membership function value larger than λ, it is considered as comparatively satisfactory. Consequently, there are four levels to judge an obtained performance, as shown in Table 4.1. With this classification, the designer may set a positive number m to require more than m performances out of the n specifications to reach at least the comparatively satisfactory level ($m \leq n$).

According to the above analysis, the final selection rule is defined as follows. Considering an analog circuit sizing problem with n constraints and m performances which are required to reach at least the comparatively satisfactory level,

(1) Given two solutions with $min(\mu_i(x)) = 0, i = 1, 2, \ldots, n$, select the solution with the smaller constraint violation;
(2) Given one solution with $min(\mu_i(x)) = 0, i = 1, 2, \ldots, n$, and one solution with $min(\mu_i(x)) > 0$ select the solution with $min(\mu_i(x)) > 0$;
(3) Given two solutions with $min(\mu_i(x)) > 0, i = 1, 2, \ldots, n$, compute the number of constraints with $min(\mu_i(x)) \geq \lambda$ (noted as $\bar{m}_1, \bar{m}_2, \bar{m}_A = min(\bar{m}_1, \bar{m}_2)$, $\bar{m}_B = max(\bar{m}_1, \bar{m}_2)$)

    (3.1) if $\bar{m}_A > m$, select the solution with the better objective function;
    (3.2) if $\bar{m}_B < m$, select the solution with the better $\sum_{i=1}^{n} \mu_i(x)$;
    (3.3) if $\bar{m}_A < m \leq \bar{m}_B$, select the solution with more constraints that reach the comparatively satisfactory level.

It can be seen that the selection mechanism achieves the four designer's intentions described above:

(1) The tolerance is restricted to the neighborhood of the specifications, because of the constructed fuzzy sets and the constraint violation-based selection rule.
(2) Only when a better objective function value appears, the candidate with lower $\sum_{i=1}^{n} \mu_i(x)$ can be selected.
(3) A sizing result with many performances near the value of $\mu(x) = 0$ that is considered as an unsatisfactory case, will not occur. The reason is that if the number of satisfactory performances is not high enough, the mechanism will select results with better $\sum_{i=1}^{n} \mu_i(x)$.
(4) Besides considering objective function values of infeasible solutions, $\sum_{i=1}^{n} \mu_i(x)$ is also optimized when the number of performances in the λ-cut set is not enough. This corresponds to the intention of the designer wishing the sizing to meet his / her expected specifications as much as possible.

4.5 Single-Objective Fuzzy Analog IC Sizing

4.5.1 Fuzzy Selection-Based Differential Evolution Algorithm

Combining DE with the above new selection mechanism, the fuzzy selection-based differential evolution (FSBDE) algorithm can be constructed by replacing the original selection mechanism in DE with the newly presented fuzzy selection mechanism. The flow diagram is shown in Fig. 4.2.

The algorithm consists of the following steps:

Step 1: Initialize the population by randomly selecting initial values of the decision variables within the allowed ranges.

Step 2: Evaluate the objective values of all individuals, and select that with the best objective value X_{best}.

Step 3: Perform the DE mutation operation for each individual to obtain each individual's mutant counterpart.

Step 4: Perform the DE crossover operation between each individual and its corresponding mutant counterpart to obtain each individual's trial individual.

Step 5: Evaluate the objective values of the trial individuals.

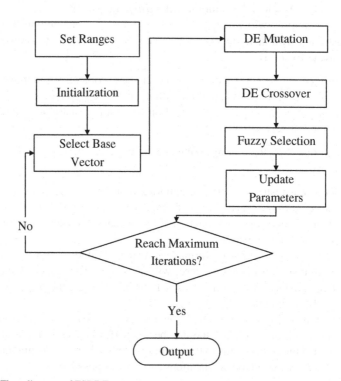

Fig. 4.2 Flow diagram of FSBDE

Step 6: Perform the fuzzy selection operation between each individual and its corresponding trial counterpart according to the selection rule described above, so as to generate the new individual for the next iteration.

Step 7: Determine the best individual of the current new population with the best objective value. If the objective value is better than the objective value of X_{best}, then update X_{best} and its objective value with the value and objective value of the current best individual.

Step 8: If the stopping criterion is met (e.g., a convergence criterion or a maximum number of iterations), then output X_{best} and its objective value; otherwise go back to Step 2.

4.5.2 Experimental Results and Comparisons

The two-stage telescopic cascade amplifier in Fig. 4.3 is used to show the application of FSBDE. The load capacitance is 1.6 pF, and the technology used is a 0.25 μm CMOS process with 2.5 V power supply. The design variables are the transistor widths (after appropriate matching conditions were enforced): W1, W1C, W3, W3C, W13, W5, W7, W9, W11, Wbn, with a range of 0.5 μm to 1000 μm; the compensation capacitance (Cc), with a range of 0.1 pF to 50 pF; the bias current (ibb), with a range of 0.5 μA to 10 mA; and the bias voltage ($vvcp$, $vvcn$), with a range of 0.05 V to 1.1 V. The transistor lengths are set to 0.25 μm. The DE step size F is 1, the

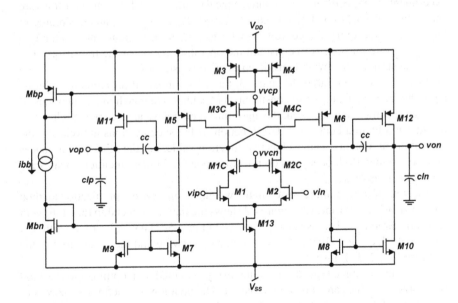

Fig. 4.3 CMOS two-stage telescopic cascode amplifier

Table 4.2 Typical results obtained by crisp SBDE with high specifications

Specifications	Constraints	Run 1	Run 2	Run 3
DC gain (dB)	≥ 70	70.28	70.03	70.17
GBW (MHz)	≥ 250	251.20	314.01	250.32
Phase margin (°)	≥ 60	60	60.06	60.12
Output swing (V)	≥ 4	4.06	4.06	4.06
Area (μm^2)	≤ 3600	3470.4	3600	3429
dm1	≥ 1.2	8.98	4.41	4.73
dm3	≥ 1.2	2.14	1.74	2.25
dm5	≥ 1.2	11.81	5.64	12.32
dm7	≥ 1.2	2.58	2.85	2.56
dm9	≥ 1.2	3.63	5.60	3.55
dm11	≥ 1.2	9.89	5.10	10.40
dm13	≥ 1.2	5.23	8.05	1.61
Power (mW)	objective	15.48	51.71	14.90
Time (s)	N.A.	1887	1829	1906

crossover probability is 0.8, and the population size is 60. All the examples were run on a Pentium IV PC at 3GHz with 2GB RAM and the Linux operating system, in the MATLAB environment.

Comparison between the results by the crisp SBDE [19], the fuzzy weighted-sum method from [1, 10] in conjunction with DE, and the FSBDE method without and with λ-cut have been carried out. Twenty random runs are performed for each experiment. The specifications and three typical results of the different algorithms are shown in Tables 4.2, 4.3, 4.4, 4.5 and 4.6. The so-called "dm" parameters represent the value of the drain-source voltage with respect to the drain-source saturation voltage, whose purpose is to constrain the operating region of the transistors. The "dm" constraints are always considered as crisp constraints.

As can be seen in Tables 4.2 and 4.3, the selection-based differential evolution algorithm can handle the crisp constraints, but cannot integrate human intentions. If the high specifications are used, the algorithm cannot have an acceptable relaxation in order to obtain a better objective function. If the specifications are relaxed, the algorithm can obtain a better objective function value, but the other performances may be too low. Both behaviors do not correspond to human intentions, who want something in between.

Two sets of fuzzy numbers used in the weighted-sum fuzzy optimization method are shown in Figs. 4.4 and 4.5. Figure 4.4 shows the fuzzy membership functions with relatively large p_j, whereas small p_j is used in Fig. 4.5. The weights of the different constraints: DC gain, GBW, phase margin, output swing and the square root of the area, are 1, 4, 2, 3, 1, respectively.

The results of three typical runs of the fuzzy weighted-sum optimization method with large p_j values are shown in Table 4.4. The notation of $\widehat{c}$ defines a fuzzy constraint (e.g., DC gain $\geq \widehat{70}$ dB). It can be seen that the results are totally unsatisfactory.

Table 4.3 Typical results obtained by crisp SBDE with low specifications

Specifications	Constraints	Run 1	Run 2	Run 3
DC gain (dB)	≥ 65	66.07	65.01	65.07
GBW (MHz)	≥ 220	226.94	222.91	220.93
Phase margin (°)	≥ 30	31.66	33.50	30.98
Output swing (V)	≥ 3	4.32	4.68	4.73
Area (μm^2)	≤ 4900	3901.3	4511.4	4137.2
dm1	≥ 1.2	4.83	5.44	6.50
dm3	≥ 1.2	1.68	1.99	1.81
dm5	≥ 1.2	13.70	12.01	13.98
dm7	≥ 1.2	2.89	6.39	7.53
dm9	≥ 1.2	5.65	17.32	21.74
dm11	≥ 1.2	10.62	14.42	16.04
dm13	≥ 1.2	6.15	8.31	4.61
Power (mW)	objective	4.97	6.30	6.31
Time (s)	N.A.	1699	1652	1646

Because of the large p_j, "too soft performances" appear, such as the GBW in solution 3. The membership function value is an acceptable 0.43 for 176.33 MHz in Fig. 4.4 b, but it is far from the specification of 250 MHz. Clearly, the p_j for the GBW also seems not large enough for the first and the second result. The membership function values of 0.006 and 5.66 are very close to 0. It can be seen that the evolution of GBW failed, because no solution with a large membership function has been found. Therefore, the algorithm tried to find larger membership function values for other performances, such as phase margin, to obtain a better weighted-sum result of the sum of the membership functions of all constraints.

A new experiment by using the same constraints with the fuzzy numbers shown in Fig. 4.5 yields no solution: the membership function values of most constraints are very close to 0. The reason is that at the initial steps of the optimization, the performances have difficulty to reach the lower bounds. If the lower bounds are relatively moderate (this can be achieved by using large p_j), the membership function values of those constraints are distinctly larger than 0 and can be enhanced if better solutions are found (see the previous experiment). If the weighted-sum method is used with tight lower bounds (small p_j), the only information that can be obtained is that all the memberships are approximately 0, so the selection cannot be executed effectively. However, if the fuzzy selection-based methods are used, although the membership function values are all approximately 0 in the first steps, the violation of the constraints are used to select better individuals.

The fuzzy numbers used in FSBDE with and without λ-cut are shown in Fig. 4.6. Notice that this is an example of fuzzy numbers used in analog IC sizing, and that different designers can use different fuzzy numbers according to their own interests.

The results of FSBDE without λ-cut shown in Table 4.5 are much better than the crisp constrained optimization algorithm and the weighted-sum method.

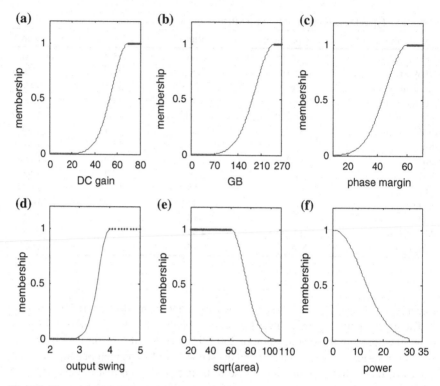

Fig. 4.4 Fuzzy membership functions with large p_j

FSBDE without λ-cut obtains better objective function values and less degradations compared to Table 4.3. The method generates solutions with good objective function and the degradation on the other performances can be accepted by the designer, such as solution 3. However, the method also generates "embarrassing" solutions, where although the objective function is good, many constraints are near the lower bounds, such as solution 2. The reason is that although the fuzzy number from the designer reflects his or her intentions, there is no distinction of "marginally accepted degradation" and "relatively satisfactory degradation". A designer cannot allow too many "marginally accepted degradation". This cannot be controlled when not using λ-cuts.

The final experiment is the full FSBDE algorithm with λ-cuts, as shown in Table 4.6. The λ and m values are determined by the analog designer according to his specific needs. In this example, we require at least 4 of the 5 performances to reach at least a comparatively satisfactory level. The satisfactory levels are 67 dB, 240 MHz, 50°, 3.5 V, 4096 μm² for DC gain, GBW, phase margin, output swing and area, respectively. Here, λ can be decided by combining the values of satisfactory levels and the membership functions. Apart from directly defining λ, this is an alternative way. Three typical results are shown in Table 4.6. It can be seen that the results

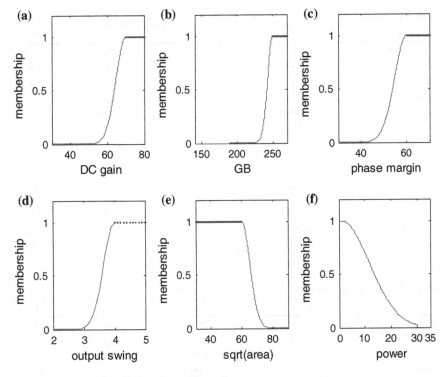

Fig. 4.5 Fuzzy membership functions with small p_j

of the FSBDE algorithm conform to the four designer's intentions summarized in Sect. 4.4. No performance has excessive relaxation. In all the results, at least four of the performance constraints reach the relatively satisfactory level. The objective function is competitive with the results of any other method in this example.

From the above experiments, it can be seen that thanks to the proposed fuzzy selection-based constraint handling method, and the differential evolution algorithm, FSBDE combines the human's flexibility and the computer's high optimization ability to obtain an intelligent automatic analog circuit sizing methodology.

4.6 Multi-objective Fuzzy Analog Sizing

Sections 4.4–4.5 discusses the single-objective analog circuit sizing with fuzzy constraints. This section will discuss multi-objective analog circuit sizing with fuzzy constraints. Recall that in Chap. 2, the NSGA-II algorithm and the selection-based method have been integrated to handle constrained multi-objective analog IC sizing

Table 4.4 Typical results obtained by weighted-sum fuzzy optimization with large p_j

Specifications	Constraints	Run 1	Run 2	Run 3
DC gain (dB)	$\geq \widehat{70}$	54.41	38.85	56.05
GBW (MHz)	$\geq \widehat{250}$	0.006	5.66	176.33
Phase margin (°)	$\geq \widehat{60}$	88.88	23.28	0.47
Output swing (V)	$\geq \widehat{4}$	2.78	2.49	2.75
Area (μm^2)	$\leq \widehat{3600}$	4243.9	44680	1201.2
dm1	≥ 1.2	10.44	2.02	12.42
dm3	≥ 1.2	6.13	1.31	1.83
dm5	≥ 1.2	1.83	1.68	1.63
dm7	≥ 1.2	2.46	2.55	2.47
dm9	≥ 1.2	1.64	2.29	1.56
dm11	≥ 1.2	1.60	1.76	3.84
dm13	≥ 1.2	19.96	1.59	9.55
Power (mW)	objective	2.67	72.99	3.22
Time (s)	N.A.	1201	1237	1073

Table 4.5 Typical results obtained by FSBDE without λ-cut

Specifications	Constraints	Run 1	Run 2	Run 3
DC gain (dB)	$\geq \widehat{70}$	75.11	75.31	83.44
GBW (MHz)	$\geq \widehat{250}$	240.68	230.13	280.23
Phase margin (°)	$\geq \widehat{60}$	45.50	48.89	49.68
Output swing (V)	$\geq \widehat{4}$	4.60	3.21	4.30
Area (μm^2)	$\leq \widehat{3600}$	2279.9	1204.7	1576.8
dm1	≥ 1.2	2.64	1.85	5.90
dm3	≥ 1.2	5.21	6.06	6.80
dm5	≥ 1.2	8.92	5.08	12.70
dm7	≥ 1.2	6.81	3.63	3.14
dm9	≥ 1.2	17.45	7.05	5.68
dm11	≥ 1.2	9.61	3.94	9.57
dm13	≥ 1.2	5.86	11.24	11.72
Power (mW)	objective	3.99	3.67	4.39
Time (s)	N.A.	1628	1683	1719

problem. In this section, fuzzy constraints will be considered and the fuzzy selection rules for dominance-based MOEA will be discussed.

4.6.1 Multi-objective Fuzzy Selection Rules

Different from single-objective optimization, where the quality of a solution is decided by the objective function value when the constraints are satisfied, here we

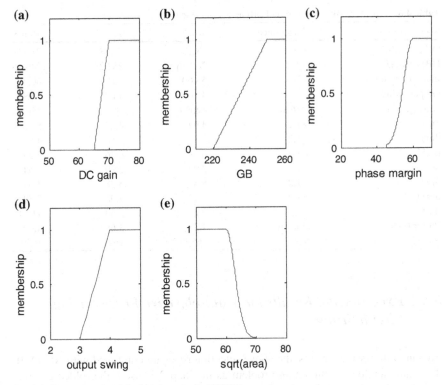

Fig. 4.6 Fuzzy membership functions in the FSBDE experiments

compare the dominance of the different objective values to decide on the quality of a candidate solution. Hence, the selection rules are modified as follows:

(1) Given two solutions with $min(\mu_i(x)) = 0, i = 1, 2, \ldots, n$, select the solution with the smaller constraint violation;
(2) Given one solution with $min(\mu_i(x)) = 0, i = 1, 2, \ldots, n$, and one solution with $min(\mu_i(x)) > 0$ select the solution with $min(\mu_i(x)) > 0$;
(3) Given two solutions with $min(\mu_i(x)) > 0, i = 1, 2, \ldots, n$, compute the number of constraints with $min(\mu_i(x)) \geq \lambda$ (noted as $\bar{m}_1, \bar{m}_2, \bar{m}_A = min(\bar{m}_1, \bar{m}_2)$, $\bar{m}_B = max(\bar{m}_1, \bar{m}_2)$)

    (3.1) if $\bar{m}_A > m$, select the solution with the better dominance. If the two solutions do not have a dominance relationship (non-dominated among them), select the one with better crowding distance;
    (3.2) if $\bar{m}_B < m$, select the solution with the better $\sum_{i=1}^{n} \mu_i(x)$;
    (3.3) if $\bar{m}_A < m \leq \bar{m}_B$, select the solution with more constraints that reach the comparatively satisfactory level.

Because the key idea of the above selection rules is the same as in single-objective constrained optimization as described before, the details will not be repeated again.

Table 4.6 Typical results obtained by FSBDE without λ-cut

Specifications	Constraints	Run 1	Run 2	Run 3
DC gain (dB)	$\geq \widehat{70}$	78.19	74.38	75.85
GBW (MHz)	$\geq \widehat{250}$	240.36	257.28	249.45
Phase margin (°)	$\geq \widehat{60}$	53.60	50.94	60.79
Output swing (V)	$\geq \widehat{4}$	4.10	4.02	3.99
Area (μm^2)	$\leq \widehat{3600}$	1203.6	1162.2	3503.8
dm1	≥ 1.2	4.92	4.22	5.26
dm3	≥ 1.2	2.80	1.58	4.83
dm5	≥ 1.2	6.58	6.36	9.21
dm7	≥ 1.2	2.92	2.99	2.54
dm9	≥ 1.2	5.57	5.28	3.63
dm11	≥ 1.2	5.57	4.98	7.79
dm13	≥ 1.2	9.22	6.69	11.51
Power (mW)	objective	4.25	3.68	4.90
Time (s)	N.A.	1650	1681	1659

4.6.2 Experimental Results for Multi-objective Fuzzy Analog Circuit Sizing

To handle fuzzy constraints, the above selection rules are combined with NSGA-II. The same analog circuit sizing problem as in Chap. 2 is used: the folded-cascode amplifier from Fig. 4.7. The comparison with the selection-based method using crisp constraints in Chap. 2 is performed. The population size is 200, and 20 runs with independent different random numbers in the NSGA-II optimization are performed for each experiment.

Then, the fuzzy selection mechanism is applied. The fuzzy membership functions used are shown by solid lines in Fig. 4.8. The third column of Table 4.7 shows the satisfactory levels of the specifications. The following two columns show a low specification and a high specification, respectively. The generated PFs using the low/high specifications serve as references for the PFs generated by MOFSS. Notice that this is an example of fuzzy numbers used in sizing, and that different designers can use different fuzzy numbers according to their own preferences.

The specifications and the average values of the obtained solutions in one typical run are shown in Table 4.8. Parameter m is set respectively to 3, 4 and 5 for the above 5 fuzzy constraints on performances, which means that at least m of the 5 constraints must meet at least the comparatively satisfactory value.

It can be seen from Fig. 4.9 that the advantage of the fuzzy selection mechanism is obvious. For example, when $m = 5$, all the specifications reach the comparatively satisfactory value, which are very near to the high specifications, but the Pareto front is much better than that of the high specifications. When $m = 3$, we can see that the Pareto front is comparable to that of using the low specifications. But at least three out of the five specifications reached the comparatively satisfactory level.

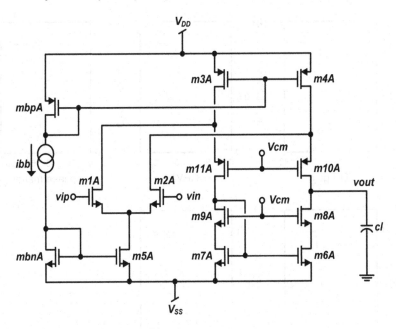

Fig. 4.7 CMOS folded-cascode amplifier

Table 4.7 Specifications used in experiments

Specifications	Fuzzy Specs	Satisfactory Level	High Specs	Low Specs
DC gain (dB)	$\geq \widehat{60}$	58	≥ 60	≥ 55
GBW (MHz)	$\geq \widehat{50}$	46	≥ 50	≥ 40
Phase margin (°)	$\geq \widehat{60}$	57	≥ 60	≥ 60
Output swing (V)	$\geq \widehat{1.2}$	1.1	≥ 1.2	≥ 1
Slew rate ($V/\mu s$)	$\geq \widehat{30}$	26	≥ 30	≥ 20
dm1	≥ 1	N.A.	≥ 1	≥ 1
dm3	≥ 1	N.A.	≥ 1	≥ 1
dm5	≥ 1	N.A.	≥ 1	≥ 1
dm7	≥ 1	N.A.	≥ 1	≥ 1
dm9	≥ 1	N.A.	≥ 1	≥ 1
dm11	≥ 1	N.A.	≥ 1	≥ 1

The relaxations reflect the designer's intention by the fuzzy sets they constructed and the comparatively satisfactory value as well as the m value they set. To sum up, the presented fuzzy selection mechanism can automatically recognize the useful relaxation and control the degree of relaxation according to the designer's intentions, so the local trade-offs can be explored successfully in one sizing process, and no additional specification tuning is necessary.

We now show how the designer controls the local trade-off based on his/her intentions by revising the fuzzy sets. In the past experiments, we assumed that the GBW

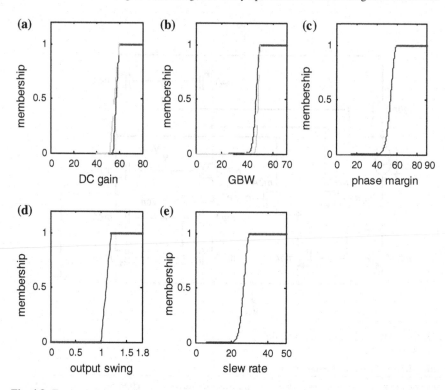

Fig. 4.8 Fuzzy membership functions in the experiments

has a comparatively large acceptable region and the DC gain has a smaller one. But if another designer has the opposite idea, as shown by the dotted line in Fig. 4.8, he/she can revise the fuzzy numbers easily to direct the multi-objective sizing. The comparatively satisfactory value is set to 53 dB for the DC gain and 46 MHz for the GBW. Parameter m is set to 4. The experiment result of a typical run (average of the 200 points in the Pareto front) is:

$$
\begin{aligned}
\text{DC gain} &= 51.21\,\text{dB} \\
\text{gain-bandwidth product} &= 46.04\,\text{MHz} \\
\text{phase margin} &= 85.16° \\
\text{output swing} &= 1.13\,\text{V} \\
\text{slew rate} &= 30.21\,\text{V}/\mu\text{s}
\end{aligned}
\tag{4.4}
$$

When we compare the DC gain and the GBW with Table 4.8, we can see that the revision of the fuzzy numbers has directed the sizing.

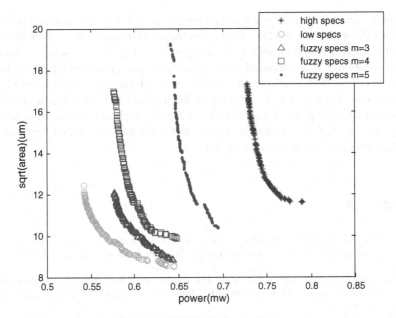

Fig. 4.9 Pareto front curves in the MOFSS experiments

Table 4.8 Experimental results with fuzzy constraints

Specifications	Constraints	Average 1	Average 2	Average 3
m	N.A.	3	4	5
DC gain (dB)	$\geq \widehat{60}$	57.08	58.13	58.12
GBW (MHz)	$\geq \widehat{50}$	42.03	42.01	46.03
Phase margin (°)	$\geq \widehat{60}$	85.43	82.45	83.51
Output swing (V)	$\geq \widehat{1.2}$	1.24	1.18	1.23
Slew rate ($V/\mu s$)	$\geq \widehat{30}$	28.30	26.25	30.05
dm1	≥ 1	17.51	17.66	17.65
dm3	≥ 1	1.63	2.03	1.94
dm5	≥ 1	5.38	3.67	3.88
dm7	≥ 1	2.53	1.66	2.16
dm9	≥ 1	1.37	1.41	1.43
dm11	≥ 1	8.72	9.10	8.34
Time (s)	N.A.	1761	1805	1836

4.7 Summary

A new kind of constraint, fuzzy constraint, has been introduced in this chapter, which
is useful in many design optimization problems. The fuzzy selection-based constraint
handling mechanism has presented a solution for problems with such constraints.
Its combination with DE and NSGA-II has generated single- and multi-objective

analog IC sizing methods. From the experimental results, it has been shown that the human's flexibility and the EA's high optimization ability have been integrated to obtain intelligent automatic analog IC sizing methodologies. The useful relaxation of specifications can automatically be recognized by the method and the designer's intentions are correctly reflected by controlling the degree of relaxation. By using fuzzy analog circuit sizing, the designer avoids tedious manual tunings of the crisp specifications to improve the circuit performance, which is achieved by a single fuzzy optimization. This technique is very useful when investigating a new circuit configuration and also in practical design.

References

1. Sahu B, Dutta A (2002) Automatic synthesis of CMOS operational amplifiers: a fuzzy optimization approach. In: Proceedings of Asia and South Pacific design automation conference, pp 366–371
2. De Smedt B, Gielen G (2003) WATSON: design space boundary exploration and model generation for analog and RF IC design. IEEE Trans Comput Aided Des Integr Circuits Syst 22(2):213–224
3. De Bernardinis F, Sangiovanni-Vincentelli A (2004) A methodology for system-level analog design space exploration. In: Proceedings of the conference on design, automation and test in Europe, pp 676–677
4. Tiwary S, Tiwary P, Rutenbar R (2006) Generation of yield-aware pareto surfaces for hierarchical circuit design space exploration. In: Proceedings of the 43rd annual design automation conference, pp 31–36
5. Eeckelaert T, McConaghy T, Gielen G (2005) Efficient multiobjective synthesis of analog circuits using hierarchical pareto-optimal performance hypersurfaces. In: Proceedings of the conference on design, automation and test, pp 1070–1075
6. Stehr G, Graeb H, Antreich K (2007) Analog performance space exploration by normal-boundary intersection and by Fourier-Motzkin elimination. IEEE Trans Comput Aided Des Integr Circuits Syst 26(10):1733–1748
7. Mueller D, Graeb H, Schlichtmann U (2007) Trade-off design of analog circuits using goal attainment and wave front sequential quadratic programming. In: Proceedings of design, automation and test in Europe conference and Exhibition, pp 1–6
8. Rutenbar R, Gielen G, Roychowdhury J (2007) Hierarchical modeling, optimization, and synthesis for system-level analog and RF designs. Proc IEEE 95(3):640–669
9. Castro-López R, Roca E, Fernández F (2009) Multimode pareto fronts for design of reconfigurable analogue circuits. Electron Lett 45(2):95–96
10. Fares M, Kaminska B (1995) FPAD: A fuzzy nonlinear programming approach to analog circuit design. IEEE Trans Comput Aided Des Integr Circuits Syst 14(7):785–793
11. Oltean G, Miron C, Zahan S, Gordan M (2000) A fuzzy optimization method for CMOS operational amplifier design. In: Proceedings of the 5th seminar on neural network applications in electrical engineering, pp 152–157
12. Torralba A, Chavez J, Franquelo L (1996a) Circuit performance modeling by means of fuzzy logic. IEEE Trans Comput Aided Des Integr Circuits Syst 15(11):1391–1398
13. Torralba A, Chavez J, Franquelo L (1996b) FASY: a fuzzy-logic based tool for analog synthesis. IEEE Trans Comput Aided Des Integr Circuits Syst 15(7):705–715
14. Coello C, Lamont G, Veldhuizen D (2007) Evolutionary algorithms for solving multi-objective problems. Springer, New York

15. Liu B, Fernandez F, Gao P, Gielen G (2009a) A fuzzy selection based constraint handling method for multi-objective optimization of analog cells. In: Proceedings of European conference on circuit theory and design, pp 611–614
16. Liu B, Fernández F, Gielen G (2009b) Fuzzy selection based differential evolution algorithm for analog cell sizing capturing imprecise human intentions. In: Proceedings of IEEE congress on evolutionary computation, pp 622–629
17. Deb K (2000) An efficient constraint handling method for genetic algorithms. Comput Methods Appl Mech Eng 186(2):311–338
18. Ross T (1995) Fuzzy logic with engineering applications. Wiley, Sussex, UK
19. Zielinski K, Laur R (2006) Constrained single-objective optimization using differential evolution. In: Proceedings of IEEE congress on evolutionary computation, pp 223–230

Chapter 5
Process Variation-Aware Analog Circuit Sizing: Uncertain Optimization

From Chaps. 2–4, this book has focused on analog circuit sizing in nominal conditions, in which the targeted problems are deterministic optimization problems. In this chapter and the next one, we concentrate on analog circuit sizing considering process variations, in which the targeted problems are uncertain optimization problems. The techniques in the first part of the book can also be applied here, and a new challenge is introduced for uncertain optimization, the trade-off of the accuracy and the efficiency. This chapter provides a general overview of uncertain optimization techniques for process variation-aware analog IC sizing problem. The next chapter will provide examples of ordinal optimization (OO)-based Monte-Carlo (MC) methods.

This chapter is organized as follows. Section 5.1 gives the introduction and problem formulation. Some traditional methods in the analog circuit design automation field are also reviewed. Section 5.2 provides a general overview of uncertain optimization. Sections 5.3 and 5.4 introduces some efficiency enhancement techniques that are common to most strategies for uncertain optimization. Section 5.5 summarizes this chapter.

5.1 Introduction to Analog Circuit Sizing Considering Process Variations

5.1.1 Why Process Variations Need to be Taken into Account in Analog Circuit Sizing

Analog integrated circuit design in industrial environments not only calls for fully optimized nominal design solutions, but also requires high robustness and yield in the light of varying supply voltage and temperature conditions, as well as inter-die and intra-die process variations [1–3]. Supply voltage variations, temperature conditions, aging of the devices and process variations are the main factors that deviate the performance of an analog circuit from its desired performance in nominal conditions.

B. Liu et al., *Automated Design of Analog and High-frequency Circuits*,
Studies in Computational Intelligence 501, DOI: 10.1007/978-3-642-39162-0_5,
© Springer-Verlag Berlin Heidelberg 2014

In nanometer CMOS technologies, random and systematic process variations have a large influence on the quality and yield of the manufactured analog circuits, and efficient and effective methods for this issue are needed. As a consequence, in the high-performance analog and mixed-signal design flows, the designer needs guidelines and tools to deal with the factors impacting circuit yield and performances in an integrated manner in order to avoid costly re-design iterations [4].

5.1.2 Yield Optimization, Yield Estimation and Variation-Aware Sizing

Yield optimization includes system-level hierarchical optimization [5] and building-block-level yield optimization [6, 7]. At the building block level, there exist parametric yield optimization [6–8] and layout-related yield optimization [9–11], e.g., critical area yield analysis [9]. Parametric yield addresses the effect of statistical variations of the design and fabrication model parameters on the circuit performance. Hence, parametric yield optimization is carried out in the sizing of the analog circuit design flow (see Chap. 1), which is different from layout-related yield optimization, which often uses a determined design and optimize the layout. This book focuses on parametric yield optimization at the building-block level.

The aim of yield optimization is to find a circuit design point x^* that has a maximum yield, considering the performance specifications and the manufacturing and environmental variations [8]. In the following, we will elaborate the design space D, process parameter space S with distribution $pdf(s)$, environmental parameter space Θ and specifications P.

The design space D is the search space formed by the circuit design points, x, which can be transistor widths and lengths, resistances, capacitances and bias voltages and currents. Each one has an upper and lower bound, which is determined by the technological process or the user's setup. The process parameter space S is the space of statistical parameters reflecting the process fluctuations, e.g., oxide thickness T_{ox} and threshold voltage V_{th}. Process parameter variations can be inter-die or intra-die. Inter-die variations are variations of the same parameter in different fabricated dies and intra-die variations are often called mismatch in analog circuit design, which are the variations between specific devices (e.g., a differential pair) within the same die. For an accurate sizing, both types should be considered. The environmental variables Θ include temperature and power supply voltage. The specifications P are the requirements set by the designer, which can be classified into performance constraints (e.g., DC gain $>70\,\mathrm{dB}$) and functional constraints (e.g., transistors must work in the saturation region). The parametric yield optimization problem can be formulated as to find a design point x^* that maximizes yield [12]:

$$x^* = arg_{x \in D} max\{Y(x)\} \tag{5.1}$$

Parametric yield is defined as the percentage of manufactured circuits that meet all the specifications considering all process and environmental variations. Hence, yield can be formulated as:

$$Y(x) = E\{YS(x, s, \theta)|pdf(s)\} \tag{5.2}$$

where E is the expected value. $YS(x, s, \theta)$ is equal to 1 if the performance of x meets all the specifications considering s (process fluctuations) and θ (environmental variations); otherwise, $YS(x, s, \theta)$ is equal to 0. Hence, the expected value of the samples can be used as the estimated yield value for a design. In most analog circuits, circuit performances change monotonically with the environmental variables θ. Then, the impact of environmental variations can be handled by simulations at the extreme values of the environmental variables. For instance, if the power supply may experience some variations, e.g, 10 %, the largest degradation is obtained by simulating at the extreme values: $(1 \pm 10\%) \times nominal\ value$. Process variations, on the other hand, are much more complex: directly simulating the extreme values (classical worst-case analysis [1]) may cause serious over-design. This book therefore focuses on the impact of statistical process variations (space S) in yield optimization.

From the concept of yield optimization, it is easy to see that the investigation of the solution methods of (5.2) is the research issue of yield estimation. When the yield optimization problem in (5.1) can be solved, it can be combined with the analog circuit sizing solutions discussed in part one of this book. Yield provides constraints to all the specifications of interest, but only optimizing yield is a constraint satisfaction problem. Can we include objective optimization into (5.1)? The problem can then be defined as minimizing some function f (e.g., power, area) subject to a minimum yield requirement y [13]:

$$\begin{aligned} x^* &= arg_{x \in D}min\{f(x, s, \theta)\}, s \in S, \theta \in \Theta \\ s.t.\ \ &Y(x) \geq y \end{aligned} \tag{5.3}$$

Equation (5.3) is the basic formula of yield- or variation-aware sizing. When (5.3) is solved, it is still interesting to investigate whether it can be combined with the multi-objective optimization in part one, as illustrated by (5.4) and (5.5).

$$x^* = arg_{x \in D}min \begin{cases} f_1(x, s, \theta) \\ f_2(x, s, \theta) \\ \vdots \\ f_n(x, s, \theta) \\ s \in S, \theta \in \Theta \\ s.t.\ Y(x) \geq y \end{cases} \tag{5.4}$$

$$x^* = arg_{x \in D}min \begin{cases} f_1(x, s, \theta) \\ -Y(x) \\ s \in S, \theta \in \Theta \\ s.t. \ Y(x) \geq y \end{cases} \tag{5.5}$$

Equation (5.4) shows a traditional multi-objective analog IC sizing method with a minimum yield constraint. Equation (5.5) is more interesting: its goal is to make a trade-off between yield and some critical performance. In this case, yield is not only a constraint, but also an objective. Because yield provides constraints to all the specifications of interest, and the designers are often not interested in a trade-off between all or most of the specifications and yield, he or she may be interested in one or two critical performances and may want to make a trade-off between these and the circuit yield. From the view of uncertain optimization, (5.4) is an expected value-based constrained uncertain multi-objective optimization problem, and (5.5) is a chance constraint-based uncertain multi-objective optimization problem. They can transform into each other and can be solved in a similar way.

5.1.3 Traditional Methods for Yield Optimization

The yield optimization flow is summarized in Fig. 5.1. In the optimization loop, the candidate circuit parameters are generated by the optimization engine; the performances and yield are analyzed and fed back to the optimization engine for the next iteration. Yield analysis is a critical point in the yield optimization flow. Among the factors that impact yield, statistical inter-die and intra-die process variations play a vital role [8].

The most straightforward way for yield optimization is to use sufficient samples to estimate the yield for each candidate solution, and follow a canonical EA framework. This is called Monte-Carlo (MC)-based methods. MC-based methods have the advantages of generality and high accuracy [14], so they are the most reliable and commonly used technique for yield estimation. Nevertheless, a large number of simulations are needed for MC analysis, therefore preventing its use within an iterative yield optimization loop (Fig. 5.1). Some speed enhancement techniques for MC simulations based on Design of Experiments (DOE) techniques have been proposed, such as the Latin Hypercube Sampling (LHS) method [13, 15] or the Quasi-Monte-Carlo (QMC) method [16, 17], to replace the Primitive Monte-Carlo (PMC) simulation. These speed improvements are significant, but if only DOE methods are used, the computational load is still too large for use in yield optimization in real practice.

Efficient yield optimization methods include device model corner-based methods [3, 18], performance-specific worst-case design (PSWCD) methods [6, 7], response-surface-based methods [2, 19] and ordinal optimization (OO)-based MC methods [20, 21]. The first two kinds of methods are traditional mathematical methods, which are reviewed in the following, while the latter two will be introduced later on.

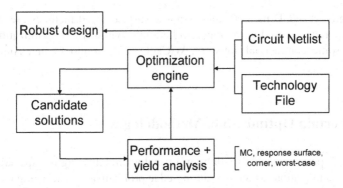

Fig. 5.1 General flow of yield optimization methods

- Device model corner-based methods [3, 18] use the same slow/fast parameter sets to decide the worst-case parameters for all circuits for a given technology. They are efficient due to the limited number of simulations needed. But their drawback is that the worst-case performance values are pessimistic as the corners correspond to the tails of the joint probability density function of the parameters, resulting in considerable over-design (power, area). Also, the slow/fast values obtained for a single performance, e.g., delay, and the worst-case process parameters for other performances may be different. On the other hand, the actual yield may be low if the intra-die variations are ignored. If the intra-die variations are considered, the number of simulations can be extremely large [1]. The limitations of device model corner-based methods for robust analog IC sizing are discussed in [1, 12].
- The PSWCD methods [6, 7, 22] represent an important progress in robust sizing of analog ICs. Instead of using the same slow/fast parameter sets for all the circuits, the PSWCD methods decide on the worst-case (WC) parameters for specific performances of each circuit and nominal design solution. Determining the performance-specific worst-case parameters is critical for this kind of method. Although the search for the WC point typically uses some nonlinear optimization formulation, most PSWCD methods [12] linearize the performances at the worst-case point, which can introduce inherent errors. Some PSWCD methods build a response surface between the inter-die parameters and the performances [22] (RSM PSWCD). The inter-die parameters are independent of the design parameters, but intra-die variations have correlations to the design parameters. Hence, intra-die variations cannot be considered in these methods. If intra-die variations are considered, the total number of process variation variables increases significantly with the number of the devices. While some PSWCD methods calculate an approximate estimation of the yield, others do not calculate yield. Instead, they calculate a range of the process parameters for a given yield, in which the specifications are met. In this case, the estimated yield is not available explicitly and the method has to be run repeatedly with different target values (e.g., $yield > 95$–99%) to know the best yield that can be achieved.

Currently, PSWCD methods and response-surface-based methods are the most popular approaches used in the repeated iterations within yield optimization loops (such as some commercial software, WiCkeD [23], Solido Design Automation Tool [24]).

5.2 Uncertain Optimization Methodologies

From the CI point of view, variation-aware analog circuit sizing is a special kind of uncertain optimization, which is a cutting-edge problem and is attracting increasing attention in recent years. Some important subareas in the uncertain optimization research are listed below [25].

- Noisy optimization
  Noisy optimization focuses on correctly identifying the optimal solutions in uncertain environments. In industrial applications, such as in system control, a high noise level exists for some problems. Due to the noise, the estimated fitness of a solution may appear either better or worse than its true fitness. The selected solutions may not necessarily be the truly good ones in the evolving population. Detrimental impacts of noise observed by Beyer [26] include the reduction of the convergence rate and premature convergence to sub-optimal solutions [25]. Therefore, the goal of the developed EAs is to avoid the negative effects caused by noises.
- Dynamic optimization
  Dynamic optimization focuses on efficiently solving problems including time varying components. In many real-world problems, the environment is changing with time, e.g., scheduling problem. Therefore, the fitness landscape also changes with time. A simple approach for these problems is to restart the optimization once a time-dependent landscape change is detected. However, this approach is too inefficient, especially when the new optimal solution set is somewhat similar to the previous solutions. Therefore, the goal of the developed EAs is to exploit past information to improve tracking performance.
- Robust optimization
  Robust optimization focuses on finding solutions that provides satisfactory performance considering parametric variations, i.e., it is insensitive to small variations in design and/or environment variables [25]. Robust optimization is widely applied in design optimization fields, since variation is unavoidable in the manufacturing processes.

It can be seen that process variation-aware analog IC sizing is a special kind of robust optimization. Its characteristics are: (1) an explicit and accurate measurement of the robustness (yield) is required; (2) efficiency enhancement is a critical problem. Note that many robust optimization methods do not have a quantitative specification of the robustness. The robustness is estimated and is traded off with the objective(s) optimality. For example, a typical solution method for robust optimization is proposed in [25]. It uses robust measure (e.g., worst case measure) to reflect the degree of variation to the objectives due to the process variations. The selection criterion adopted is randomly based on either the conventional Pareto ranking or the robust measure to make a balance between the Pareto-optimality and the robustness. The goal is to generate a set of robust Pareto-optimal designs. However, for the targeted problem, the robustness is measured as the yield value, which is an explicit constraint or objective. Yield ($Y(x)$) represents the probability of satisfying all the specifications, and the requirement of $Y(x) \geq \theta$ is called a chance constraint [27]. Chance constraints have been widely used in many real-world applications such as mechanical engineering [28], electrical engineering, reliability engineering [29].

To measure the robustness to the process variations, which are stochastic variables with certain distributions, the most general and accurate technique is MC simulation. However, the drawback is that the solution of yield optimization and variation-aware analog circuit sizing are computationally expensive due to the MC simulations. In many industrial applications, the time taken to perform one single simulation is on the order of a few seconds to minutes. Although the simulation time of each sample is not high, MC simulations are needed to evaluate the objectives and constraints, which cost a much longer time. The total number of evaluations that can be performed is limited by the time constraint: a few days optimization time can often be tolerated, but several weeks or more usually is not acceptable to meet the time-to-market requirements. Hence, it is of practical interest to reduce the number of simulations. Unlike part one of this thesis, where enhancing the optimization and constraint handling ability is the main goal, efficiency enhancement is the main goal in this part. The objective is to reduce the number of simulations but to retain a high optimization ability.

It is obvious that this task is not trivial. For the yield optimization and process variation-aware analog circuit sizing problems, some additional challenges exist:

- There are often tens of decision variables in these problems, which are given by the characterization of modern CMOS technologies. The traditional off-line surrogate model-based methods for uncertain optimization [30] are limited by the dimensionality of the problem. In these methods, surrogate models of the yield over the designable and process parameters are first established through regression methods and these are subsequently used to estimate the yield in the sizing process. To obtain a reliable surrogate model, a sufficient amount of well spread training data points needs to be generated first. The trained surrogate model can then replace the real function evaluation in the optimization process. For example, [31] uses artificial neural networks (ANN) [32] to train an offline surrogate model to the performances of the system. However, the necessary number of points to cover the design space increases exponentially with the number of decision variables. When the dimension of the problem becomes larger, the process of generating the training data is often very expensive. Yet, many areas that provide infeasible or inferior solutions in the decision space do not contribute much to the optimization, so the simulations of those points are wasted. Also, these methods suffer from the trade-off between the accuracy and the complexity of the model, as well as between the accuracy and the number of samples (CPU time) to create the model.

- Advanced MC sampling methods can help but is still not fast enough. Currently, most approaches for uncertain optimization still use primitive Monte-Carlo (PMC) simulation for each candidate solution to obtain the expected values of the performances of the system, and then combine this with conventional EAs. To reduce the computational cost, some reported approaches use Design of Experiments (DOE) methods, e.g., LHS, to replace PMC simulation [33]. QMC method was introduced in [16, 17], which shows better performance compared to LHS. Besides basic DOE methods, other state-of-the-art MC simulation methods have been applied, e.g., Shuffled Complex Evolution Metropolis, which is a Markov chain Monte-Carlo (MCMC) method, was introduced in [34] to perform the samplings in noisy uncertain optimization problems. Although these approaches have introduced important advances, they still have limitations. LHS often cannot provide a sufficient speed enhancement with respect to PMC [17]. MCMC-based advanced sampling methods are often sensitive to some parameters or settings (e.g., start distribution), which require problem-specific knowledge [17, 35]. Convergence diagnostic is very critical in MCMC-based approaches, but there are few general and robust methods to determine how many steps are needed to converge to the stationary distribution within an acceptable error [36]. The QMC method is difficult to solve high-dimensional problems [37], but tens of process variation parameters are common for practical analog ICs.

Two kinds of CI-based methods show good performances for the targeted problem. They are response-surface-based methods [38] and ordinal optimization (OO)-based MC methods [20, 21]. The response-surface-based methods in [38] is in fact a surrogate model assisted evolutionary algorithm (SAEA). A surrogate model is constructed on-line using the simulated data, which is used to predict the performances of new candidate solutions. Verifications with different cost are used. Full MC simulation is only performed for the promising candidates after comparatively cheap evaluations. Using response-surface-based methods (RSM) to solve variation-aware analog IC sizing problems is detailed in [39]. RSM methods show good potential for the targeted problem. There are three important factors involved in the development of an RSM: (1) How to build the surrogate model; (2) How to handle the prediction uncertainty of the surrogate model; (3) How to integrate the surrogate model with evolutionary algorithms. They will be elaborated in part three of this book for another application, high-frequency IC design automation. The OO-based MC method based on the intelligent allocation of samples according to the quality of the candidate solutions will be elaborated in the next chapter.

5.3 The Pruning Method

As said above, a key method to improve the efficiency is to reduce the number of MC simulations. There are two possible approaches to achieve this goal: (1) pruning unnecessary MC simulations, (2) using fewer MC simulations to achieve the same estimation accuracy. Two efficiency improvement techniques that can almost be combined with all MC-based uncertain optimization methods will be introduced.

This section takes yield as an example to illustrate the pruning method, but it is applicable to all uncertain optimization problems with chance constraint. In the yield optimization process, some candidate solutions may appear that cannot satisfy the specifications even for nominal values of the process parameters. Their yield values would be too low to become a useful candidate solution. Hence, there is not much sense in applying the MC-based yield estimation to these solutions. We can simply assign them a zero yield value without any MC simulation. Their violation of the constraints in nominal condition can be calculated, and the constrained optimization algorithm will minimize the constraint violations in order to move the search space to feasible solutions (i.e., design points that satisfy the specifications for nominal process parameters).

We can further develop the pruning method when different evaluation methods with different cost exist. Take analog IC for example. DC and AC analysis are quite cheap, and transient analysis is more expensive (but still much cheaper than MC simulation). In this way, for candidate solutions with low quality results for DC and AC analysis, no transient analysis is needed. An example is shown in [38].

5.4 Advanced MC Sampling Methods

MC simulation is needed to evaluate candidate solutions for most uncertain optimization problems because of its advantages of generality and high accuracy [16, 17]. Hence, a key efficiency enhancement idea is to get good estimation accuracy with as few samples as possible. To address this problem, advanced MC simulation method has been intensively investigated and is a key component of computational statistics. In this section, we take the yield estimation of analog circuit as an example to review some advanced MC simulation techniques.

The estimation of yield is an approximation of the integral of the function that determines if the design specifications are met with the statistical process parameters varying over the unit cube [1]:

$$Y = \int, \dots, \int_{f \in A_f} pdf(f) \cdots df, df = df_1 \cdots df_2, \dots, df_n \qquad (5.6)$$

where A_f is the acceptance region. The integration error can be separated into the factor related to the function itself and the factor related to the generated set of random points according to the Koksma-Hlawka theorem [40], which is shown as:

$$|\widehat{I} - I| \leq D_n^*(x_1, x_2, \dots, x_n) V_{HK}(f) \qquad (5.7)$$

where $\widehat{I}$ is the estimated value of the integration; I is the real value of the integration; D_n^* is the star discrepancy which measures the uniformity of the generated points:

more uniformly distributed samples have a lower D_n^*; $V_{HK}(f)$ is the total variation of f [41], which is determined by the problem itself and n is the number of samples.

The separation of the difference between the estimated and real values of the integration in two components, D_n^* and $V_{HK}(f)$, in Eq. (5.7) leads to 2 categories of advanced MC simulation methods and their usage in circuit yield estimation:

- Variance-reduction methods (e.g., importance sampling (IS) [1, 42], LHS [15]), which focus on decreasing $V_{HK}(f)$.

  In IS the good shifted distribution function is often circuit specific, which poses the challenge of generality. LHS [43, 44] first creates equal slices in each dimension of the stochastic variable vector, and then selects random values within each slice for every coordinate. At last, by randomly matching up the coordinate values, a bunch of LHS samples are constructed. Because of this stratification technique, the variance is reduced. LHS is a general method, but the performance is not always good enough, especially for some problems that are difficult to be decomposed into a sum of univariate functions [37]. LHS has been shown to perform worse than Quasi-Monte-Carlo (QMC) [16, 17] in some circuit examples.

- Low-discrepancy sequence (LDS)-based methods (e.g., QMC [16, 17]), which focus on decreasing D_n^*.

  QMC is a general method for different problems. It uses LDS to generate more uniformly distributed samples, which is shown to have considerable asymptotic advantage compared to PMC. LDS is a deterministic infinite sequence of d-dimensional points. The goal of the constructed sequences is to fill the space as geometrically equidistant as possible. However, its major drawback is that for high-dimensional problems, the asymptotic advantage of the QMC point set appears to require an impractically large number of samples to set [37]. For instance, a 50-stochastic-variables case is usual in yield estimation of analog circuits, but using the Sobol set with base 2 (standard), the advantage on the asymptotic rate of convergence can be expected after 2^{127} samples with the standard settings. If we use a lower number of samples, the first few dimensions are sampled uniformly, but higher dimensions are not, which degrades the performance. To address this problem, [16, 17] present a method, whose core idea is to sacrifice the non-uniformity in higher dimensions, but making the loss as small as possible. This is done by first ranking the sensitivities of the process variation variables and by selecting the important variables that mainly dominate the variance. They are mapped to the first few dimensions of LDS, which are more uniformly distributed. We can call this method "ranking-integrated QMC". The theoretical background of this method is analysis of variance (ANOVA) decomposition (a method for decomposing variance [45]) and the concept of effective dimensions [46]. Experiments show a significant progress compared with PMC and LHS for medium-scaled (i.e., tens of stochastic variables) yield estimation problems [16]. However, even with this method, some important variation variables still have to use non-uniformly distributed samples in many cases. The reason is that the upper limit on dimensions for LDS to keep the uniformity is typically 10–12 [47] if a reasonable number of samples can be used to maintain the efficiency in QMC. But the number of important stochastic variables (or effective dimensions) is typically larger than this threshold. Hence, a question still remains: What is the better solution when the number of effective dimensions (or important variables that dominate the variance, which is problem dependent) is larger than the upper limit on dimensions for LDS (i.e., higher than say 12)? A solution is the Latin Supercube Sampling (LSS) method, which is appropriate for candidate solution fitness estimation in uncertain optimization.

5.4.1 AYLeSS: A Fast Yield Estimation Method for Analog IC

Analog yield estimation using Latin Supercube Sampling (AYLeSS) was proposed in [48] and has demonstrated good performance for practical analog circuits. Although AYLeSS takes analog circuit yield estimation as an example, it can be used for a wide range of problems with MC simulations involving many stochastic variables.

5.4.1.1 Basics of QMC

Because QMC is a part of the AYLeSS method, QMC is briefly introduced first. In PMC, the uniformly distributed "random numbers" generated by computers are not truly uniform, and the gaps arising among the samples adversely affect the uniformity. QMC, on the other hand, constructs a deterministic infinite sequence of d-dimensional points and selects a certain number of them when performing sampling. The goal of the constructed sequences is to fill the space as geometrically equidistant as possible. Such sequences are called low-discrepancy sequences. If the integrand has a bounded variation in the sense of Hardy and Krause [37], it is possible to construct a LDS with asymptotic rate:

$$D_n^* = O(n^{-1}(logn)^d) \tag{5.8}$$

where n is the number of samples, d is the number of process variation variables and D_n^* is the star discrepancy (see Chap. 5). This proves that an asymptotic rate better than that of PMC, which has a rate of $n^{-1/2}$ [49], can be achieved. There are different methods to generate a LDS, e.g., Halton set, Sobol set, etc. In AYLeSS, the Sobol set with a skip of $2^{\lfloor logn/log2 \rfloor}$ points is used, because a LDS often has better performance when skipping the first few points [50]. The details of constructing the Sobol set can be found in [41].

5.4.1.2 The Partitioning Method in AYLeSS

To be able to handle high-dimensional problems, different from ranking-integrated QMC, the main idea of AYLeSS is not to accept the degradation of QMC at high dimensions and minimize the loss, but to decrease the number of dimensions to avoid such loss. The method consists in partitioning the high-dimensional problem into sub-groups with lower dimensions. In this way, generating a high-dimensional LDS is transformed to generating some groups of low-dimensional LDS, and, in each group, the uniformity can be kept with a reasonable number of samples if the dimensionality of each group is not large.

However, the partitioning neglects the interactions between different sub-groups. Hence, the partitioning technique may affect the result. The best partitioning technique is to arrange variables that interact more strongly into a sub-group, and the interactions between different sub-groups should be as small as possible. The reason is that the interactions are considered within a sub-group, but not between separate sub-groups. This rule is quite intuitive. Suppose we divide the d dimensions into k groups, then each group has s dimensions, where $d = k \times s$. In the extreme case, different sub-groups have no interaction with each other; then the error is the sum of the errors in each sub-group. Hence, we can expect that the variance converges at the rate of a s-dimensional problem. Now, the problem becomes how to construct a set of easy to implement rules to obtain a good partitioning. In AYLeSS, the rules are constructed based both on the aspects of analog circuit design and statistics, as follows:

(1) The dimension of each sub-group should not be larger than 12 [47], because if the dimension of the sub-group is too large, the uniformity of LDS will also be sacrificed.
(2) It is not wise to use too many sub-groups, because the fewer the number of sub-groups, the less interactions between different sub-groups need to be considered.
(3) It is better that the number of dimensions of each sub-group is as similar as possible, because the convergence rate is often determined by the sub-group with the highest dimension.
(4) Devices whose widths and lengths have symmetry correlations (e.g., differential pairs) or have clear design relations interacting strongly (e.g., current mirror) should be clustered into one sub-group when process variations are considered.
(5) The intra-die variables of interacting transistors should be clustered into one sub-group.

According to these rules, the following heuristic partitioning method can be constructed:

Step 0: Include the statistical inter-die variables into one sub-group. If the dimension of the sub-group is larger than 12, partition them into 2 or more groups. The dimensions of each group should be as similar as possible.

Step 1: Find differential pairs and current mirrors in the circuit.

Step 2: Include the intra-die variables of each differential pair and current mirror into one cluster.

Step 3: Combine the clusters with small dimensions from Step 2 to larger sub-groups whose dimensions should not be larger than 12. If the dimension of any cluster is larger than 12, split it into 2 or more sub-groups with dimensions less than or equal

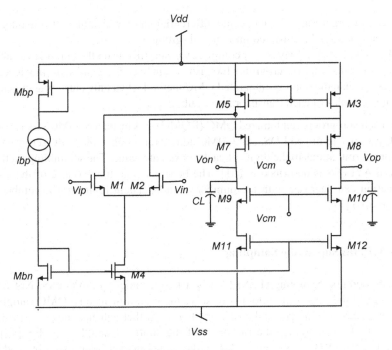

Fig. 5.2 CMOS fully differential folded-cascode amplifier

to 12, maintaining the intra-die variables of each transistor together in a sub-group. The dimensions of all subgroups should be as similar as possible.

Step 4: For all other transistors that are not in differential pairs and current mirrors, partition them into different sub-groups. The dimensions should be as similar as the sub-groups from the previous steps. The intra-die variables of each transistor should be in the same sub-group.

Let us consider as an example the circuit in Fig. 5.2 in a standard 0.35 μm CMOS technology. In this circuit, the following transistors are matched: M1-M2, M3-M5-Mbp, M7-M8, M9-M10, M4-Mbn and Mbn-M11-M12. The technology has 15 inter-die process parameters, and each transistor has 4 stochastic intra-die parameters $(V_{th}, T_{ox}, W_{eff}, L_{eff})$. According to the partitioning rules, the process variables can be divided as follows: group 1: inter-die variables (8 variables); group 2: inter-die variables (7 variables); group 3: intra-die variables of M1-M2 (8 variables); group 4: intra-die variables of M3-M5-Mbp (12 variables); group 5: intra-die variables of M7-M8 (8 variables); group 6: intra-die variables of M9-M10 (8 variables); group 7: intra-die variables of M11-M12 (8 variables); group 8: intra-die variables of Mbn-M4 (8 variables).

This partitioning method is often applicable to analog circuits because:

(1) Most technologies use 4 intra-die variables $(V_{th}, T_{ox}, W_{eff}, L_{eff})$. Even for a complex technology, the number of intra-die variables for a differential pair is

seldom larger than 12. Hence, the differential pairs, which have strong interaction, can often be clustered into one sub-group.

(2) Even if the technology is very complex, we can still cluster the intra-die variables of one transistor in a sub-group. Experiments in Sect. 5.4.2 still show much better results than the ranking-integrated QMC method when only clustering the intra-die variables of each transistor into a sub-group.

In each sub-group, randomized QMC (RQMC) sets instead of QMC sets is used. QMC uses a determined LDS, and RQMC adds scrambling to the QMC set by using random permutations to the digits of each coordinate value. The scrambling method used in AYLeSS is described in [51]. The benefit is that the variance of the yield estimation can decrease with the number of independent replications (number of sub-groups) [37].

5.4.1.3 Latin Supercube Sampling

Latin Supercube Sampling (LSS) [37] is a key technique of AYLeSS. Not only does it hold the idea of partitioning to solve the high-dimensional QMC sampling problem, LSS also integrates the LHS method to further enhance the performance. The method to randomize the run order of the stratified samples of LHS [47] is applied to the QMC sub-groups. Hence, the points in each group are obtained by randomly permuting the run order of the QMC points (the permutations of different groups are independent). Suppose x is a d-dimensional input sample. x is divided into k groups, where $d = k \times s$ (in real applications, the dimensions of the sub-groups are not necessarily equal). $x_i^j, i = 1, 2, \ldots, n$ is an s-dimensional QMC point set ($j = 1, 2, \ldots, k$). The ith LSS sample is:

$$xlss_i = (x_{\pi 1(i)}^1, x_{\pi 2(i)}^2, \ldots, x_{\pi k(i)}^k), i = 1, 2, \ldots, n \qquad (5.9)$$

where πj are independent uniform random permutations of $1, 2, \ldots, n$. The purpose of this random permutation is the same as in LHS, i.e., to make the projection of each coordinate of the samples more uniform so as to reduce the variance (now coordinate refers to the sub-group). It has been proven that even with no partitioning rules, a poor grouping can still be expected to do equally well as LHS [37]. Therefore, the convergence rate of the AYLeSS method can always be better than LHS for complex and high-dimensional problems.

5.4.1.4 Building AYLeSS

In summary, the AYLeSS method works as follows:

Input: d-dimensions of the variation variables x, sample size n, joint probability of the variation variables $\Pi(x)$.

Step 0: Skip $2^{\lfloor \log n / \log 2 \rfloor}$ points of the Sobol sequence.

Step 1: Partition the input variation variables into k groups with dimensions $\{s_1, s_2, \ldots, s_k\}$ according to the partitioning rules.

For EACH sub-group:

Step 2: Scramble the Sobol set according to the method in [51].

Step 3: Select the RQMC points according to the dimension of the sub-group.

Step 4: Perform a random permutation to the run order of the RQMC points according to Eq. (5.9) to obtain the LSS sample X_s.

End

Step 5: Generate the required samples by $\Pi^{-1}(X_s)$ for circuit simulation, and calculate the yield.

5.4.2 Experimental Results of AYLeSS

Two designs are shown as examples with yield from 80 to 90 % for a typical analog circuit in a 90 nm technology. For each design, the comparisons to ranking-integrated QMC, pure QMC, LHS and PMC are shown. The same ranking method and the setting of 1000 initial MC samples for ranking (counted in the total number of samples) in [49] are used. For pure QMC, the LDS coordinates are assigned randomly. In data analysis, the confidence interval (e.g., 1 % error compared to the true yield value) under a certain confidence level (e.g., 2σ) is used to reflect the estimation error. Two properties, the standard deviation convergence rate and the necessary sample size to obtain a certain accuracy, are selected as the comparison criteria, which is the same as in [49]. 10 runs are performed for AYLeSS, the QMC-based methods and LHS, and 5 runs are performed for the PMC method (because the computational cost is too high for PMC). From these data, the standard deviation for different numbers of samples for each method can be obtained. The $\sigma \propto n^{-\alpha}$ relationship is assumed and the corresponding convergence exponent, α, can be estimated by linear fitting and least square error is used in the fitting. By the same fitting method, the sample size needed for each method can also be estimated for a given level of accuracy (or estimation error) using the central limit theorem [52]:

$$\sigma \Phi^{-1}\left(\frac{1+p}{2}\right) \leq Y\left(\frac{\delta}{100}\right) \tag{5.10}$$

where Φ is the standard normal cumulative distributed function, Y is the exact value of the yield (estimated by 500,000 PMC samples), δ is the confidence interval and p is the confidence level. We use a 2σ confidence level and a 1% confidence interval (the same as [49]) to compute the required σ. With σ derived from Eq. (5.10), the required sample size, N_{req}, can be estimated by the fitted function, $\sigma \propto n^{-\alpha}$.

The two-stage fully differential folded-cascode amplifier with common-mode feedback in Fig. 5.3 is designed in a 90 nm CMOS process with 1.2 V power supply. The specifications are

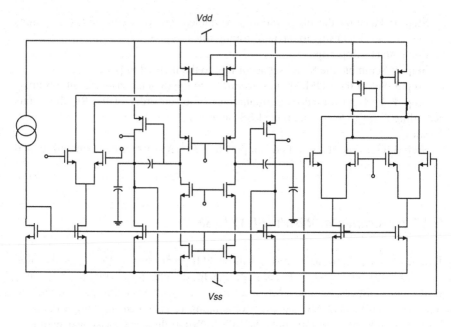

Fig. 5.3 CMOS two-stage fully differential folded-cascode amplifier

Table 5.1 Results obtained by the different methods

Design 1	N_{req}	$-\alpha$	Design 2	N_{req}	$-\alpha$
AYLeSS	1265	−0.631	AYLeSS	807	−0.6236
QMC	3681	−0.4842	QMC	2148	−0.5202
Ranking QMC	2767	−0.6816	Ranking QMC	2187	−0.5729
LHS	2448	−0.53	LHS	3687	−0.4963
PMC	6831	−0.5189	PMC	4307	−0.4904

$$
\begin{aligned}
&\text{DC gain} \geq 60\,\text{dB} \\
&\text{gain-bandwidth product} \geq 45\,\text{MHz} \\
&\text{phase margin} \geq 60° \\
&\text{output swing} \geq 1.9\,\text{V} \\
&\text{power} \leq 2.5\,\text{mW} \\
&\text{area} \leq 250\,\mu\text{m}^2
\end{aligned}
\tag{5.11}
$$

The number of design variables is 21 and the variation parameters can be extracted from 52 standard normal-distributed random numbers for the selected circuit. Two designs with a yield of 84.29 and 90.39 %, respectively, are shown as examples. Table 5.1 shows the results. The fitted lines in log10-log10 scale for the different methods are shown in Figs. 5.4 and 5.5.

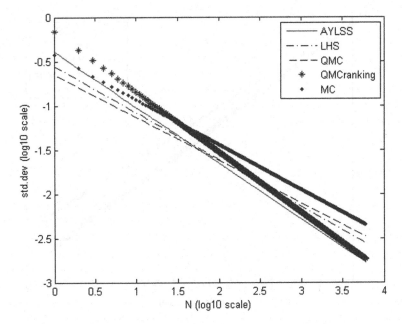

Fig. 5.4 Fitting of the convergence rates for the different methods (design 1)

From the columns with the necessary number of samples (N_{req}) in Table 5.1, it can be concluded that:

(1) The AYLeSS result is the best one compared to the other 4 methods for both designs. AYLeSS can achieve a 2–2.5 times speed enhancement compared with ranking-integrated QMC and more than a 5 times speedup compared with PMC.
(2) If we decrease the 1000 samples (used for ranking) for the ranking-integrated QMC method, we can see that the new N_{req} numbers, 1767 and 1187, are still worse than the AYLeSS results. This shows that the large number of effective dimensions degrades the QMC sampling even with ranking. If the number of effective dimensions is larger than the threshold (e.g., 12), some important dimensions cannot obtain uniformly distributed samples. In contrast, AYLeSS, does not have this problem and can often lower the scales by partitioning while receiving good results.

From the columns of the convergence exponent ($-\alpha$) in Table 5.1, it can be concluded that:

(1) The ranking is important for the QMC method, as the convergence rate considerably decreases without ranking.
(2) Designs with 85–99 % yield are the most interesting ones to obtain an accurate estimation, because they may become a real design product. However, designs with high yield need less MC samples, and the required number of samples for 50 % yield is the largest [1]. From this relatively large-scale circuit, we can get

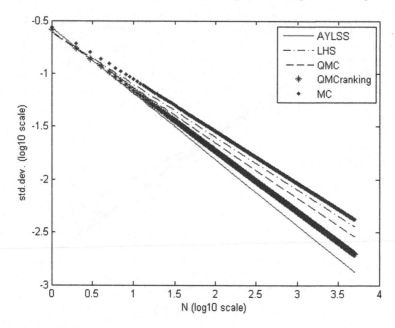

Fig. 5.5 Fitting of the convergence rates for the different methods (design 2)

a rough idea of the necessary number of samples for a yield larger than 85 %
for typical analog circuits when using LDS-based methods. Hence, the 1000
additional samples needed for QMC ranking is expensive for these cases. In
contrast, AYLeSS does not need ranking and the partitioning rules are easy to
implement.

Next, the results obtained by LSS using a simplified partitioning method are
shown. In the five partitioning rules described above, besides those related to the
LSS sampling itself (e.g., requirements on dimension), there are two rules specific
for analog circuits. In this example, we keep the intra-die variables of each transistor
in a group (rule 5), but do not consider the symmetry (matching) of devices (rule
4). The dimensions of all the sub-groups containing intra-die variables are 8. Three
different groupings are used when combining the intra-die variables of the transistors
to form sub-groups. Besides the inter-die variables, each of the above sub-groups
contains the intra-die variables of two transistors, which are picked randomly. The
results are shown in Table 5.2.

From Table 5.2, it can be concluded that:

(1) The performances in Table 5.2 are lower than that of AYLeSS in Table 5.1, i.e.,
the required number of samples is larger. Hence, the partitioning rule of clustering the symmetry and matching devices into one sub-group can enhance the
performance.

Table 5.2 LSS results for different groupings

Design 1	N_{req}	$-\alpha$	Design 2	N_{req}	$-\alpha$
LSS1	1317	−0.6596	LSS1	822	−0.4518
LSS2	1313	−0.6280	LSS2	842	−0.6132
LSS3	1450	−0.6232	LSS3	914	−0.5908

(2) Even when only clustering the intra-die variables of each transistor in a group, the result is still better than ranking-integrated QMC and LHS.

(3) Different ways of partitioning all provide good results if the rule of clustering the intra-die variables of each transistor in a group is satisfied. Hence, AYLeSS is quite robust with respect to the partitioning used.

5.5 Summary

This chapter has introduced the basics of uncertain optimization with the practical example of variation-aware analog circuit sizing. The problems of yield optimization and single/multi-objective process variation-aware sizing of analog IC have been defined. The targeted problem is a special kind of robust optimization problem in the uncertain optimization research area. The challenges encountered have been discussed, which are the high efficiency requirement and the ability to handle high-dimensional problems. At present, promising solution methods for the targeted problems are on-line response-surface-based methods and ordinal optimization-based Monte-Carlo method. Common efficiency enhancement techniques for almost all kinds of uncertain optimization algorithms, the pruning method and advanced MC simulation methods, have been introduced at last.

References

1. Graeb H (2007) Analog design centering and sizing. Springer Publishing Company, Incorporated
2. Gielen G, Eeckelaert T, Martens E, McConaghy T (2007) Automated synthesis of complex analog circuits. In: Proceedings of 18th European conference on circuit theory and design, pp 20–23
3. Eshbaugh K (1992) Generation of correlated parameters for statistical circuit simulation. IEEE Trans Comput Aided Des Integr Circuits Syst 11(10):1198–1206
4. Buhler M, Koehl J, Bickford J, Hibbeler J, Schlichtmann U, Sommer R, Pronath M, Ripp A (2006) DATE 2006 special session: DFM/DFY design for manufacturability and yield-influence of process variations in digital, analog and mixed-signal circuit design. In: Proceedings of design, automation and test in Europe, vol 1, pp 1–6
5. Yu G, Li P (2008) Yield-aware hierarchical optimization of large analog integrated circuits. In: Proceedings of IEEE/ACM international conference on computer-aided design, pp 79–84
6. Schenkel F, Pronath M, Zizala S, Schwencker R, Graeb H, Antreich K (2001) Mismatch analysis and direct yield optimization by specwise linearization and feasibility-guided search. In: Proceedings of the 38th design automation conference, pp 858–863

7. Mukherjee T, Carley L, Rutenbar R (2000) Efficient handling of operating range and manufacturing line variations in analog cell synthesis. IEEE Trans Comput Aided Des Integr Circuits Syst 19(8):825–839
8. McConaghy T, Palmers P, Gao P, Steyaert M, Gielen G (2009a) Variation-aware analog structural synthesis: a computational intelligence approach. Springer Verlag, Dordrecht, Netherlands
9. Khademsameni P, Syrzycki M (2002) Manufacturability analysis of analog CMOS ICs through examination of multiple layout solutions. In: Proceedings of 17th IEEE international symposium on defect and fault tolerance in VLSI Systems, pp 3–11
10. Xu Y, Hsiung K, Li X, Pileggi L, Boyd S (2009) Regular analog/RF integrated circuits design using optimization with recourse including ellipsoidal uncertainty. IEEE Trans Comput Aided Des Integr Circuits Syst 28(5):623–637
11. Chen J, Luo P, Wey C (2010) Placement optimization for yield improvement of switched-capacitor analog integrated circuits. IEEE Trans Comput Aided Des Integr Circuits Syst 29(2):313–318
12. Schwencker R, Schenkel F, Pronath M, Graeb H (2002) Analog circuit sizing using adaptive worst-case parameter sets. In: Proceedings of design, automation and test in Europe conference and exhibition, pp 581–585
13. Tiwary S, Tiwary P, Rutenbar R (2006) Generation of yield-aware pareto surfaces for hierarchical circuit design space exploration. In: Proceedings of the 43rd annual design automation conference, pp 31–36
14. Mutlu A, Gunther N, Rahman M (2003) Concurrent optimization of process dependent variations in different circuit performance measures. In: Proceedings of the 2003 international symposium on circuits and systems, vol 4, pp 692–695
15. Stein M (1987) Large sample properties of simulations using Latin hypercube sampling. Technometrics 29:143–151
16. Singhee A, Singhal S, Rutenbar R (2008) Practical, fast Monte Carlo statistical static timing analysis: why and how. In: Proceedings of the IEEE/ACM international conference on computer-aided design, pp 190–195
17. Singhee A, Rutenbar R (2009) Novel algorithms for fast statistical analysis of scaled circuits. Springer, Netherlands
18. Barros M, Guilherme J, Horta N (2010) Analog circuits optimization based on evolutionary computation techniques. Integr VLSI J 43(1):136–155
19. Basu S, Kommineni B, Vemuri R (2009) Variation-aware macromodeling and synthesis of analog circuits using spline center and range method and dynamically reduced design space. In: Proceedings of 22nd international conference on VLSI design, pp 433–438
20. Liu B, Fernández F, Gielen G (2011a) Efficient and accurate statistical analog yield optimization and variation-aware circuit sizing based on computational intelligence techniques. IEEE Trans Comput Aided Des Integr Circuits Syst 30(6):793–805
21. Liu B, Zhang Q, Fernández F, Gielen G (2013a) An efficient evolutionary algorithm for chance-constrained bi-objective stochastic optimization and its application to manufacturing engineering. IEEE Trans Evol Comput (To be published)
22. Sengupta M, Saxena S, Daldoss L, Kramer G, Minehane S, Cheng J (2005) Application-specific worst case corners using response surfaces and statistical models. IEEE Trans Comput Aided Des Integr Circuits Syst 24(9):1372–1380
23. MunEDA (2013) MunEDA homepage. http://www.muneda.com/index.php
24. Solido (2013) Solido Design Automation homepage. http://www.solidodesign.com/
25. Goh C, Tan K (2009) Evolutionary multi-objective optimization in uncertain environments: issues and algorithms. Springer, Berlin
26. Beyer H (2000) Evolutionary algorithms in noisy environments: theoretical issues and guidelines for practice. Comput Methods Appl Mech Eng 186(2):239–267
27. Liu B (2010) Uncertain programming. Uncertainty theory, pp 81–113
28. Mercado LL, Kuo SM, Lee TY, Lee R (2005) Analysis of RF MEMS switch packaging process for yield improvement. IEEE Trans Adv Packag 28(1):134–141
29. Chan H, Englert P (2001) Accelerated stress testing handbook. IEEE Press, New York

30. Liu B (2002) Theory and practice of uncertain programming. Physica Verlag, Heidelberg
31. Lv P, Chang P (2008) Rough programming and its application to production planning. In: Proceedings of international conference on risk management & engineering management, pp 136–140
32. Wasserman P (1989) Neural computing: theory and practice. Van Nostrand Reinhold Co., New York
33. Zhang Q, Liou J, McMacken J, Thomson J, Layman P (2001) Development of robust interconnect model based on design of experiments and multiobjective optimization. IEEE Trans Electron Dev 48(9):1885–1891
34. Nazemi A, Yao X, Chan A (2006) Extracting a set of robust Pareto-optimal parameters for hydrologic models using NSGA-II and SCEM. In: Proceedings of IEEE congress on evolutionary computation, pp 1901–1908
35. Andrieu C, De Freitas N, Doucet A, Jordan M (2003) An introduction to MCMC for machine learning. Mach Learn 50(1):5–43
36. Cowles M, Carlin B (1996) Markov chain Monte Carlo convergence diagnostics: a comparative review. J Am Stat Assoc 91:883–904
37. Owen A (1998) Latin supercube sampling for very high-dimensional simulations. ACM Trans Model Comput Simul (TOMACS) 8(1):71–102
38. McConaghy T, Gielen G (2009) Globally reliable variation-aware sizing of analog integrated circuits via response surfaces and structural homotopy. IEEE Trans Comput Aided Des Integr Circuits Syst 28(11):1627–1640
39. McConaghy T, Palmers P, Steyaert M, Gielen G (2009b) Variation-aware structural synthesis of analog circuits via hierarchical building blocks and structural homotopy. IEEE Trans Comput Aided Des Integr Circuits Syst 28(9):1281–1294
40. Hlawka E (1961) Funktionen von beschränkter variatiou in der theorie der gleichverteilung. Annali di Matematica Pura ed Applicata 54(1):325–333
41. Hickernell F (1998) A generalized discrepancy and quadrature error bound. Math Comput 67(221):299–322
42. Doorn T, Ter Maten E, Croon J, Di Bucchianico A, Wittich O (2008) Importance sampling Monte Carlo simulations for accurate estimation of SRAM yield. In: Proceedings of 34th European solid-state circuits conference, pp 230–233
43. Keramat M, Kielbasa R (1997) Latin hypercube sampling Monte Carlo estimation of average quality index for integrated circuits. Analog Integr Circuits Signal Process 14(1):131–142
44. Dharchoudhury A, Kang S (1993) Performance-constrained worst-case variability minimization of VLSI circuits. In: Proceedings of 30th conference on design automation, pp 154–158
45. Rackwitz R (2002) Optimization and risk acceptability based on the life quality index. Struct Saf 24(2):297–331
46. Caflisch R, Morokoff W, Owen A (1997) Valuation of mortgage backed securities using Brownian bridges to reduce effective dimension. Department of Mathematics, University of California, Los Angeles, Technical Report
47. McKay M, Beckman R, Conover W (1979) A comparison of three methods for selecting values of input variables in the analysis of output from a computer code. Technometrics 21:239–245
48. Liu B, Messaoudi J, Gielen G (2012c) A fast analog circuit yield estimation method for medium and high dimensional problems. In: Proceedings of the conference on design, automation and test in Europe, pp 751–756
49. Singhee A, Rutenbar R (2010) Why Quasi-Monte Carlo is better than Monte Carlo or Latin hypercube sampling for statistical circuit analysis. IEEE Trans Comput Aided Des Integr Circuits Syst 29(11):1763–1776
50. Acworth P, Broadie M, Glasserman P (1998) A comparison of some Monte Carlo and quasi-Monte Carlo techniques for option pricing. In: Monte Carlo and Quasi-Monte Carlo methods in scientific computing, vol 127, pp 1–18
51. Matousek J (1998) On the L2-discrepancy for anchored boxes. J Complex 14(4):527–556
52. Fischer H (2011) A history of the central limit theorem: from classical to modern probability theory. Springer, New York

Chapter 6
Ordinal Optimization-Based Methods for Efficient Variation-Aware Analog IC Sizing

As was discussed in Chap. 5, variation-aware analog circuit sizing is a kind of robust optimization problem in the uncertain optimization research area. To guarantee that variation-aware analog circuit sizing can be finished in a practical time, the efficiency of the optimization process is specially emphasized for this kind of problem, although, of course, high-quality solutions are also required. For the problems from Chaps. 1–4, the evaluation of a candidate design only needs a single SPICE simulation, which is computationally cheap, and the time cost to finish the sizing is often not a limitation. In contrast, in variation-aware analog IC sizing, a large number of samples are needed to evaluate the yield of a candidate design, which is computationally expensive. Besides the pruning methods and the advanced Monte-Carlo (MC) simulation methods introduced in Chap. 5 which are applicable to most uncertain optimization mechanisms, on-line response-surface based methods and ordinal optimization (OO)-based MC methods are promising mechanisms for the targeted problem. The former method has been introduced in detail in [1], and this chapter will introduce the OO-based MC methods.

The remainder of the chapter is organized as follows. Section 6.1 introduces the ordinal optimization technique, to intelligently allocate simulation budgets. Section 6.2 introduces the efficiency improvement methods based on evolutionary operators. Section 6.3 focuses on the method to scale up a CI technique integrating OO with an evolutionary algorithm for yield optimization. An example is provided. Experimental results are provided for the presented method, called OO-based random-scale differential evolution (ORDE) algorithm in Sect. 6.4. Section 6.5 discusses how to extend the yield optimization to single-objective variation-aware analog IC sizing. Section 6.6 introduces a state-of-the-art method for multi-objective variation-aware analog IC sizing based on OO. Note that these methods use analog ICs as an example, but they are applicable to general problems considering yield as well as other uncertain performance metrics. Section 6.7 summarizes this chapter.

B. Liu et al., *Automated Design of Analog and High-frequency Circuits*,
Studies in Computational Intelligence 501, DOI: 10.1007/978-3-642-39162-0_6,
© Springer-Verlag Berlin Heidelberg 2014

6.1 Ordinal Optimization

OO has emerged as an efficient technique for selecting the best candidate in a group of candidates with random variations, especially for problems where the computation of the simulation models is time consuming [2]. OO is based on two basic tenets:

- Obtaining the "order" of a set of candidates is easier than estimating an accurate "value" of each candidate. The convergence rate of ordinal optimization is exponential. This means that the probability that the observed best solution is among the true best solutions grows as $O(e^{-\alpha n})$ where α is a positive real number and n is the number of simulations [2]. In contrast, the convergence rate of methods aimed at estimating the right value instead of the order, e.g., the direct Monte-Carlo method (primitive MC), is at most $O(1/\sqrt{n})$ [3].
- An accurate estimation is very costly but a satisfactory value can be obtained much easier.

Therefore, OO meets the objective of correct selection with a reasonably accurate yield estimation and with the smallest computational effort. According to OO, a large portion of the simulations should be conducted with the critical solutions in order to reduce their estimation variance. On the other hand, limited computational effort should be spent on non-critical solutions that have little effect on identifying the good solutions, even if they have large variances. This leads to the core problem in ordinal optimization: how to allocate the computing budget. This can be formulated as follows: given a pre-defined computing budget, how should it be distributed among the candidate designs?

Consider the yield evaluation function. For a single simulation (e.g., a sample of process parameters), we define $YS(x, s) = 1$ if all the circuit specifications are met, and $YS(x, s) = 0$ otherwise. Recall that x represents the design variables and s represents the process variation variables. Because the MC simulation determines the yield as the ratio of the number of functional chips to all fabricated chips, the mean value of $YS(x, s)$, corresponds to the yield value, $Y(x)$. Let us consider a total computing budget equal to TO simulations. Using the pruning method introduced in Chap. 5, TO can be determined by the number of feasible solutions (i.e., solutions that meet the performance constraints for nominal values of the process parameters) at each iteration. Here, we set $TO = sim_{ave} \times M1$, where $M1$ is the number of feasible solutions and sim_{ave} is the average budget (number of simulations) for each candidate set by the user. The budget allocation problem consists in determining the number of simulations $n_1, n_2, \ldots, n_{M1}$ of the $M1$ candidate solutions such that $n_1 + n_2 + \ldots + n_{M1} = TO$. For this problem, several algorithms have been reported in the specialized literature [4, 5]. An asymptotic solution to this optimal computing budget allocation problem is proposed in [4]:

$$n_b = \sigma_b \Big(\sum_{i=1, i \neq b}^{M1} n_i^2 / \sigma_i^2 \Big)^{1/2}$$

$$n_i / n_j = \left(\frac{\sigma_i / \delta_{b,i}}{\sigma_j / \delta_{b,j}} \right)^2 \tag{6.1}$$

$$i, j \in \{1, 2, \ldots, M1\}, i \neq j \neq b$$

where b is the best design[1] of the $M1$ candidate solutions. For each candidate solution, some samples are allocated. For each sample, the corresponding $YS(x, s)$ can be computed (0 or 1). By these $YS(x, s)$, we can calculate their mean (estimated yield, $Y(x)$) and $\sigma_1^2, \sigma_2^2, \ldots, \sigma_{M1}^2$, which are the finite variances of the $M1$ solutions, respectively. They measure the accuracy of the estimation. Parameter $\delta_{b,i} = Y_b(x) - Y_i(x)$ represents the deviations of the estimated yield value of each design solution with respect to that of the best design. The interpretation of (6.1) is quite intuitive. If $\delta_{b,i}$ is large, the estimated yield value of candidate design i is bad, and according to $n_i / n_j = (\frac{\sigma_i / \delta_{b,i}}{\sigma_j / \delta_{b,j}})^2$, n_i becomes small, i.e., we should not allocate many simulations to this design. However if σ_i is large, it means that the accuracy of the yield estimation is low, and we should allocate more simulations to this design to obtain a better yield estimate. Therefore, the quotient $\sigma_i / \delta_{b,i}$ represents a trade-off between the yield value of design i and the accuracy of its estimation. Therefore, an OO-based yield analysis algorithm can be designed as follows:

Step 0: Let $k = 0$, and perform n_0 simulations for each feasible design, i.e., $n_i^k = n_0, i = 1, 2, \ldots, M1$.
Step 1: If $\sum_{i=1}^{M1} n_i^k \geq TO$, stop the OO for yield analysis.
Step 2: Consider Δ additional simulations (refer to [2] for the selection of the Δ and n_0 values and Sect. 6.4 for details in this application) and compute the new budget allocation n_i^{k+1} by (6.1). If $n_i^{k+1} \geq n_{max}$, then $n_i^{k+1} = n_{max}$.
Step 3: Perform additional $max\{0, n_i^{k+1} - n_i^k\}$ simulations for design $x_i, i = 1, 2, \ldots, M1$. Let $k = k + 1$ and go to step 1.

Parameter n_0 is the initial number of simulations for each candidate solution, selected to provide a very rough idea of the yield. More simulations are allocated according to the quality of the candidate later on. Parameter n_{max} is the upper limit on the number of simulations for any candidate. The value of n_{max} must call for a balance between the accuracy and the efficiency.

A typical population from the experiments in Sect. 6.4 is selected to show the benefits of OO (see Fig. 6.1): candidates with a yield value larger than 70 % correspond to 31 % of the population, and are assigned 56 % of the simulations. Candidates with a yield value smaller than 40 % correspond to 33 % of the population, and are only assigned 12 % of the simulations. The total number of simulations used is only 10.2 % of those of the same MC simulation method without OO applied to the same

---

[1] Best candidate design represents the candidate with the highest estimated yield value based on the available samples.

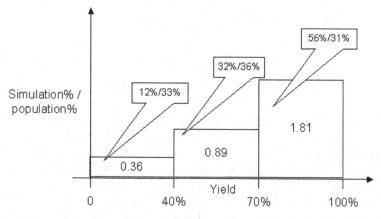

Fig. 6.1 The function of OO in a typical population

candidate designs, because repeated MC simulations of non-critical solutions are avoided.

6.2 Efficient Evolutionary Search Techniques

There are two main factors affecting the efficiency of an uncertain optimization method. They are the computational effort (number of MC simulations) in each iteration, and the necessary number of iterations to obtain the optimal solution. The pruning method, the advanced MC simulation methods and OO aim to decrease the first factor. This section introduces the methods to improve the second factor.

A main advantage of EA is to explore the search space in order to obtain globally optimal solutions, but they are not so efficient in local tuning to reach the exact optimal solution. Enhancing the ability of local exploitation of EAs can considerably reduce the necessary iterations. In the following, two methods are presented.

6.2.1 Using Memetic Algorithms

In order to enhance the local search ability, using memetic algorithms (MA) is a natural choice. Recall from Chap. 3 that, MAs [6] use, besides a global optimization engine, a population-based strategy coupled with individual search heuristics capable of performing local refinements. Two local search engines were presented, which are the simulated annealing (SA) method and the Nelder-Mead (NM) simplex method. DE is used in [7] as the global optimizer (exploration) to obtain the near-optimal solution and the NM simplex method [8] as the local search engine to refine the result provided by DE (exploitation). That method shows 1.5 times speed enhancement compared to the case where MA is not used for an analog IC yield optimization example.

Although memetic search can often enhance the solution quality and use fewer iterations to achieve the convergence, the memetic search operation itself costs additional computational effort. For example, the NM simplex method needs about 10 iterations for each candidate and each iteration needs a number of simulations. Multiplied by the size of the population, this is also an expensive procedure. Therefore, it is not wise to apply the local search operator to all the candidates of the population. Recall from Chap. 3 that, in the general framework of MA, there is a step to select individuals which undergo local refinements. This step is often omitted when the candidate solution evaluation procedure is cheap. In contrast, for the uncertain optimization problems which need a number of simulations for candidate evaluation, this step is very much needed. A possible approach is only performing local search to the best (few) member(s) of the population obtained by the global optimization engine. An additional possibility is that the local search may not be necessary in each iteration. A feasible method is to trigger the local search when the objective function value (e.g., yield) cannot be improved by the evolutionary operators for a few subsequent iterations. If this occurs, we can then use the local search operators to exploit the best member of the current population and then come back to the EA. With this approach, the memetic search method can be used in a computationally expensive MC-based optimization loop. An example is shown in [7].

6.2.2 Using Modified Evolutionary Search Operators

Standard evolutionary search operators can be modified to improve the exploitation ability. There are various methods to achieve this. For example, new crossover and mutation operators to enhance the local search ability based on DE are developed in [9–11]. As an example, the random-scale differential evolution (RSDE) [12] will be introduced in what follows.

The difference of RSDE and standard DE is the mutation operator. The new mutation operator aims to combine and balance the global and local search mechanisms. When the objective function with uncertainty (e.g., yield) is optimized to an acceptable level, local tuning is often important for further optimization. Usually, in this condition, there are many candidates which are assigned the maximum number of simulations to estimate the performance metric, which is expensive, but otherwise the accuracy would degrade very significantly. Also, the global optimization mechanism must be maintained in this period. Otherwise, the optimization has a high risk to be stuck at a premature solution. Therefore, both global and local search need to be emphasized. Before introducing RSDE, the standard DE/best/1 mutation operator is recalled:

$$V_i(t + 1) = x_{best}(t) + F(x_{r1}(t) - x_{r2}(t)) \qquad (6.2)$$

Fig. 6.2 Mutant vectors
obtained by the random-scale
operator

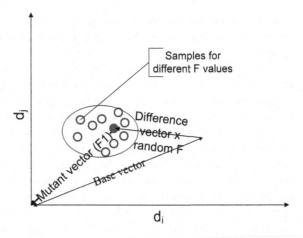

Instead of performing a separate global and local search, RSDE combines global search and local search into a unified procedure. In the original DE mutation, the scaling factor F is a constant for all the candidate solutions. If F is small, the whole evolution process is slow; if F is large, it is difficult to perform effective local fine-tunings. To solve the problem, a natural idea is to randomize F in Eq. (6.2) to each differential variation. By randomizing F, the differences of vectors can be amplified stochastically, and the diversity of the population is retained. This introduces three advantages:

(1) The algorithm has a lower probability of providing premature solutions because of the reasonable diversity;
(2) The vicinity of the mutant vector that the standard DE can be explored, which is investigated by the randomized amplification of the differential variation $x_{r1}(t) - x_{r2}(t)$. In a standard DE search process, the candidates may get stuck in a region and make the evolution quite slow when the global optimization point is nearly reached (but the diversity is also maintained). This is called "stagnation" [13]. Figure 6.2 shows the effect of randomizing F. It can be seen that a cloud of potential points centered on the mutant vector with constant scaling factor $F1$ have the potential to be investigated.
(3) Compared to MAs, local search can be widely used without the computational burden caused by a separate local tuning procedure.

For the scaling factor, it is advisable to use a vector $\widehat{F}$ composed of Gaussian-distributed random variables with mean value μ and variance σ: $\widehat{F}_{i,j} = norm(\mu, \sigma)$, $i = 1, 2, \ldots, NP$, $j = 1, 2, \ldots, d$, where NP is the population size and d is the dimensionality of the problem. A Gaussian distribution is selected based on the following two considerations:

(1) As the purpose of the random scaling factor is to search the vicinity of the mutant vectors by the constant F, it should not be far from it. By using a Gaussian distribution, 68 % of the generated samples in $\widehat{F}$ are within $\pm 1\sigma$.

(2) It should have the ability to escape from the "stagnation". A Gaussian distribution can also provide 5 % of $\widehat{F}$ values out of 2σ.

For each variable in the search space, the scaling factor $\widehat{F}_{i,j}$ of each differential variation $x_{r1}(t) - x_{r2}(t)$ is now different. Equation (6.2) is therefore changed to:

$$V_i(t + 1) = x_{best}(t) + \widehat{F}_i(x_{r1}(t) - x_{r2}(t)) \tag{6.3}$$

In Sect. 6.4, it will be shown that the necessary number of iterations of the yield optimization algorithm decreases significantly by the RSDE mechanism.

6.3 Integrating OO and Efficient Evolutionary Search

This section discusses the method to integrate OO and efficient evolutionary search mechanisms to construct an efficient analog circuit yield optimization method. Again, yield optimization is an example, and the method is applicable to most uncertain optimization problems requiring highly optimized result and high efficiency.

In the OO-based random-scale differential evolution (ORDE) algorithm [12], the main objectives are (1) to reduce the computational effort at each iteration by optimally allocating the computing budget assigned to each candidate in the population; and (2) to decrease the necessary number of iterations, hence, decrease the number of expensive MC simulations, by improving the search mechanism.

In order to optimally allocate the computing budget at each iteration, instead of assigning the same number of simulations for the MC simulations of all solutions selected by the pruning method (see Chap. 5) during the whole optimization process, the yield optimization process is divided into two stages (see Fig. 6.3). In the first stage, the ranking of the candidate solutions and a reasonably accurate yield estimation result for good (critical) solutions are important. For medium or bad (non-critical) candidate solutions, their ranking is very important, but an accurate yield estimation is not significant. The reason is that the function of the yield estimation for non-critical candidates is to guide the selection operator in the EA, but the candidates themselves are not selected as the final result nor even to enter the second stage. Hence, the computational effort spent on feasible but non-optimal candidate solutions can be strongly reduced. On the other hand, the estimations of non-critical candidates cannot be too inaccurate. After all, correct selection of candidate solutions in the yield optimization algorithm is necessary. In this first stage, the yield optimization problem is formulated as an ordinal optimization problem, targeted at identifying critical candidate solutions, allocating a sufficient number of samples to the MC simulation of these solutions, while reasonably few samples are allocated to non-critical solutions. The above technique is used until the yield converges closely to the desired value. For example, if the desired target yield is 99 %, the threshold value between the first and the second stage can be 97 %. Candidates with an estimated yield larger than the threshold value enter the second stage.

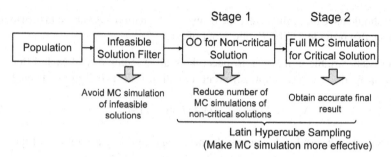

Fig. 6.3 Two-stage yield estimation flow

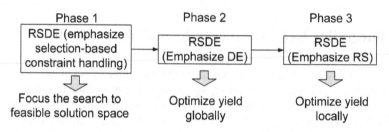

Fig. 6.4 Yield optimization flow

The threshold value must be properly selected. A too low threshold value may cause a low efficiency, as OO would stop when the yield values of the selected points are not promising enough (e.g., a 50 % yield threshold for a requirement of 90 % yield) and would shift the yield estimation and selection tasks to the second stage, which is more computationally expensive. A too high threshold value (e.g., a threshold equal to the yield requirement) may cause a low accuracy. The reason is that in most cases the points selected by OO are promising, because the candidates can be compared and selected correctly, but the estimated yield values are not sufficiently accurate for the final result. Assigning the threshold value to be two percentage points below the required target yield represents an appropriate trade-off between efficiency and accuracy.

In the second stage, an accurate result is highly important, so the number of simulations in each yield estimation increases in this stage to obtain an accurate final result and this is not as cheap as yield analysis in the first stage. Other candidates in the population that do not meet the threshold still remain in the first stage and still use the OO-based estimation method. Note that the two stages are therefore not separated sequentially in time, but rather use different yield estimation methods.

For the yield optimization flow (see Fig. 6.4), the algorithm is a selection-based random-scale differential evolution algorithm (RSDE), which is a combination of three different techniques. Each technique plays a significant role in each phase. The first phase applies a tournament selection-based method (see Chap. 2) to focus the search into the feasible solution space, defined by the nominal values of the

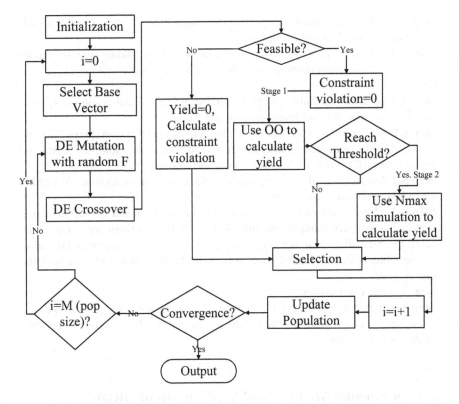

Fig. 6.5 Flow diagram of ORDE

process parameters. The goal of this phase is to prune infeasible candidates in nominal conditions and to obtain more solutions with yield value larger than 0. In the second phase, the global exploration of the DE algorithm is emphasized. The goal of this phase is to obtain candidates with reasonably good (but not the best) yield values. The random-scale operator in Sect. 6.2 is emphasized in the third phase for fine-tunings in order to obtain solutions with highly optimized yield.

Based on the above components, the overall ORDE algorithm for analog yield optimization can now be constructed. The detailed flow diagram is shown in Fig. 6.5.

The ORDE algorithm consists of the following steps.

Step 0: Initialize parameters n_0, T, Δ, n_{max} and the DE algorithm parameters (e.g., the population size NP, the crossover rate CR). Initialize the population by randomly selecting values of the design variables x within the allowed ranges.

Step 1: Update the current best candidate. If no candidate meets the specifications for nominal values of the process parameters, the best candidate is the one with the smallest constraint violation. Otherwise, the best candidate is the feasible candidate with the largest estimated yield.

Step 2: Perform the mutation operation according to Eq. (6.3) to obtain each candidate solution's mutant counterpart.

Step 3: Perform the crossover operation between each candidate solution and its corresponding mutant counterpart to obtain each individual's trial individual (see Chap. 2).

Step 4: Check the feasibility of the trial individual. For feasible solutions, go to step 5.1; for infeasible solutions, go to step 5.2.

Step 5.1: Set constraint violations equal to 0, and use the OO technique to calculate the yield. If the estimated yield is higher than the threshold value, add additional samples to perform the full MC simulation. Go to step 6.

Step 5.2: Set yield equal to 0, and calculate the constraint violations. No yield is estimated in this step. Go to step 6.

Step 6: Perform selection between each candidate solution and its corresponding trial counterpart according to the rules in [14]: if both of them are not feasible, select the one with the smaller constraint violation; if one is feasible and the other is infeasible, select the feasible one; if both are feasible, select the one with the higher yield.

Step 7: Update population.

Step 8: If the stopping criterion is met (e.g., a convergence criterion or a maximum number of generations), then output x_{best} and its objective function value; otherwise go to Step 1.

6.4 Experimental Methods and Verifications of ORDE

6.4.1 Experimental Methods for Uncertain Optimization with MC Simulations

There are several aspects that have to be considered when designing the experiments for MC simulation-based uncertain optimization methods. First, the number of MC simulations for each feasible candidate should be decided. There is not much sense in comparing the efficiency without a good accuracy. For ORDE, the number of MC simulations in the second stage is the main factor that influences the accuracy of the final result. The accuracy of the yield estimates is related to the number of samples according to [15]:

$$n_{MC} \simeq \frac{Y(1 - Y)k_\gamma^2}{\Delta Y^2} \tag{6.4}$$

where Y is the yield value and ΔY is the confidence interval. For instance, if the yield estimate is 90 %, and $\Delta Y = 1$ %, then the confidence interval is 89–91 %. Parameter k_γ reflects the confidence level, e.g., $k_\gamma = \pm 1.645$ denotes a 90 % confidence level. From Eq. (6.4), the necessary number of MC simulations can be calculated. However, this corresponds to the primitive MC (PMC) simulation. For advanced MC

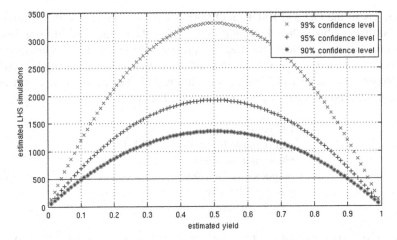

Fig. 6.6 Necessary numbers of LHS simulations as a function of the yield value and the confidence level

methods, a rough speed enhancement compared to PMC can often be estimated. For example, according to [16], LHS sampling requires 20–25 % of the number of samples compared with PMC to get a comparable accuracy. Using this assumption, Fig. 6.6 shows the estimated number of LHS simulations needed for a confidence level of 90, 95, and 99 % respectively when $\Delta Y = 1\,\%$. The number of LHS simulations is estimated as 20 % of the necessary number of PMC samples. It can be seen that even for a 99 % confidence level, for a yield larger than 96 %, 500 LHS points are sufficient. For a 90 % confidence level, 500 LHS points are even sufficient for a yield larger than 90 %. Hence, in the experiments, 500 LHS simulation is used as n_{max} (the maximum number of samples for yield estimation).

Second, the estimated yield result is influenced by the number of samples. Two experiments using 50 and 500 MC simulations for each feasible candidate can report a solution with "100 % yield", but the true yield value can be quite different. To reflect the real accuracy, we calculate the yield estimated by 50,000 LHS MC simulations at the same design point. From Eq. (6.4), we can calculate that with 99 % confidence level and $\Delta Y = 0.1\,\%$, the corresponding yield value of 50,000 LHS simulations is 96 %. When the yield of solutions waiting to be tested is higher than 96 %, an estimation result from 50,000 LHS simulations is a very reliable approximation of the real yield value for use as a reference result.

6.4.2 Experimental Verifications of ORDE

The ORDE algorithm is demonstrated with a practical analog circuit in a 90 nm CMOS technology. The MC sampling method used in ORDE is LHS. To highlight the effects of the OO and RSDE in ORDE, it will be compared with a reference method that combines the DE optimization engine, the tournament selection-based constraint handling method, the infeasible pruning (IP) and the LHS techniques.

The circuit used is a two-stage fully differential folded-cascode amplifier with common-mode feedback (CMFB), shown in Fig. 6.7. The circuit is designed in a 90 nm CMOS process with 1.2 V power supply. The specifications are

$$
\begin{aligned}
&\text{DC gain} \geq 60\,\text{dB} \\
&\text{gain-bandwidth product} \geq 45\,\text{MHz} \\
&\text{phase margin} \geq 60° \\
&\text{output swing} \geq 1.9\,\text{V} \\
&\text{power} \leq 2.5\,\text{mW} \\
&\text{area} \leq 250\,\mu\text{m}^2
\end{aligned}
\tag{6.5}
$$

There exist 21 design variables. The transistor width has a range from 0.12 to 800 μm; the transistor length has a range from 0.1 to 20 μm; the compensation capacitance has a range from 0.1 to 50 pF; the biasing current has a range of 0.05 to 50 mA. All transistors must be in the saturation region. The total number of process variation variables for this technology is 143, including 24 transistors × 4 intra-die variables/transistor = 96 intra-die variables and 47 inter-die variables. However, some of these process variation variables are correlated, and the foundry provides a method to use 52 independent random variables to obtain the values of the 143 process variation variables.

Experiments with 300 and 500 simulations for each feasible candidate by the reference IP + LHS method have been done. The improvement provided by the introduced OO technique and the improvement provided by the random-scale operator are separately studied. The results of the yield estimation provided by a 50,000 MC simulation analysis at the same final design point and the total number of simulations are analyzed. The statistical results of 10 independent runs are shown in Tables 6.1, 6.2 and Fig. 6.8.

From the best, worst and average yield values in Table 6.1, it can be seen that the accuracy with 300 simulations is obviously lower than with 500 simulations. From Fig. 6.8, we can see that the deviations of ORDE from the target value are very close to that of using 500 simulations, but the computational cost is much lower. With respect to the number of simulations, shown in Table 6.2, ORDE costs only 11.32 % of the number of simulations of the IP + LHS method with a comparable accuracy. These

Table 6.1 The yield results (using 50,000 MC simulations) of the solutions obtained by the different methods

Methods	Best (%)	Worst (%)	Average (%)	Variance (%)
300 simulations (IP + LHS)	99.0	97.9	98.3	0.002
500 simulations (IP + LHS)	99.3	98.2	98.9	0.002
OO + IP + LHS	99.7	98.1	98.9	0.003
ORDE	99.6	98.3	98.9	0.002

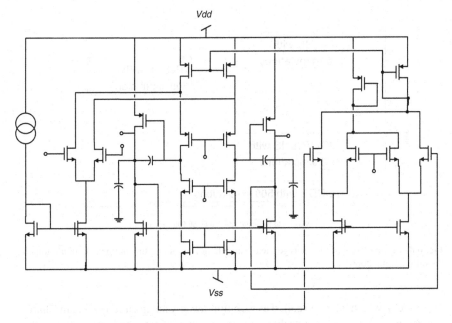

Fig. 6.7 CMOS two-stage fully differential folded-cascode amplifier

Table 6.2 Total number of simulations

Methods	Best	Worst	Average
300 simulations (IP + LHS)	1,15,500	5,46,900	2,64,130
500 simulations (IP + LHS)	1,72,500	6,88,000	4,18,730
OO + IP + LHS	39,828	1,40,537	90,209
ORDE	16,335	1,00,795	47,421

results come from the contribution of both the OO and the random-scale operator. Without the random-scale operator, as can be seen from the result of the OO + IP + LHS method, it spends 21.54 % of the simulations of the IP + LHS method. The average CPU time of ORDE for this example is 25 min. It can therefore be concluded that ORDE improves the CPU time by an order of magnitude for the same accuracy compared to the improved MC-based method integrating the infeasible pruning and Latin Hypercube sampling techniques.

6.5 From Yield Optimization to Single-Objective Analog Circuit Variation-Aware Sizing

In real practice, if the yield requirement can be met, the designers sometimes want to further optimize some objective function (e.g., power or area) while maintaining the

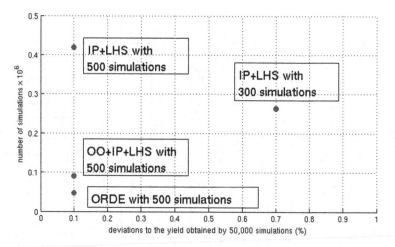

Fig. 6.8 Comparisons of the average yield estimate deviation and the number of simulations for the different methods

target yield, which has been called as variation-aware analog circuit sizing in Chap. 5. We can model the targeted problem to an uncertain constrained optimization or to an uncertain multi-objective optimization problem. This section will introduce the former one.

6.5.1 ORDE-Based Single-Objective Variation-Aware Analog Circuit Sizing

In single-objective variation-aware analog circuit sizing, both the objective function f (e.g., power) and the constraint (yield Y) must be considered simultaneously. Hence, we first look at the differences between them. Yield is not a stochastic variable, but we have some uncertainties on its estimation. If we perform an infinite number of MC simulations, yield would have an exact value. The objective function, or any performance specification, is different. If we perform an infinite number of MC simulations, power would still have a probability distribution function, but with an accurate mean and an accurate variance caused by the process variations. Therefore, for yield, it is natural to use its expected value to describe it. For the objective function, a possible way is to use the 3σ value to guarantee the reliability of the expected objective function value, where σ is extracted from the samples.

It is not difficult to extend an algorithm for yield optimization to an algorithm for single-objective variation-aware sizing. The method is to combine it with a constraint handling method. For example, we can add an outer selection procedure considering the objective function value and the yield as constraint to ORDE. The detailed selection rules are now as follows: for each candidate solution and its corresponding trial counterpart,

(1) if none of them are feasible using nominal process parameters, select the one with the smaller constraint violation;
(2) if one is feasible and the other is infeasible using nominal process parameters, select the feasible one;
(3) if both are feasible using nominal process parameters, then

    (3.1) if both of them violate the yield constraint, select the one with the smaller yield constraint violation;
    (3.2) if one satisfies the yield constraint and the other does not, select the feasible one;
    (3.3) if both of them satisfy the yield constraint, select the one with the smaller $\bar{f}(x) + 3\sigma_{f(x)}$ (f is the objective function to be minimized, $\bar{f}$ is the mean value).

Using the above selection rule to replace the original selection rule in ORDE, the extended ORDE for single-objective variation-aware analog circuit sizing can be implemented. We can roughly divide the algorithm into two phases: the yield satisfaction phase and the objective function optimization phase. If we handle the single-objective variation-aware analog circuit sizing problem as a new task, the yield satisfaction phase will be run first. However, we already have the candidates that satisfy the yield constraint since the plain yield optimization is done first to check if the yield requirement can be met. Therefore, a feasible method is to use the last population from the plain yield optimization as the initial population of the extended ORDE to prevent the yield satisfaction phase from running two times. To sum up, for the yield satisfaction phase, standard ORDE is used. The objective function optimization phase uses the final population of the standard ORDE as its initial population and then replaces the selection rules of ORDE to the new rules listed above, the single-objective variation-aware analog circuit sizing can be achieved.

6.5.2 Example

Here we use the example from the last section with the specifications of

$$
\begin{aligned}
&\text{DC gain} \geq 60\,\text{dB} \\
&\text{gain-bandwidth product} \geq 45\,\text{MHz} \\
&\text{phase margin} \geq 60° \\
&\text{output swing} \geq 1.9\,\text{V} \\
&\text{power} \leq 2.5\,\text{mW} \\
&\text{area} \leq 250\,\mu\text{m}^2 \\
&\text{settling time} \leq 25\,\mu\text{s with 1\,\%error band}
\end{aligned}
\tag{6.6}
$$

Transient analysis is needed for the settling time simulation, which is more costly than other specifications using AC simulation. The yield specification is 99 % and the power is the target design objective to be minimized. Five tests with different random seeds were performed. For plain yield optimization without optimizing the power consumption, ORDE satisfies the yield specification, 99 %, at a power of 2.38 mW. With the extended ORDE to minimize power while maintaining a yield larger than 99 %, the average power value now becomes 1.63 mW.

6.6 Bi-objective Variation-Aware Analog Circuit Sizing

Two kinds of multi-objective variation-aware analog circuit sizing problem have been defined in Chap. 5 (see (5.4) and (5.5)). In this section, we focus on the latter case. Not only does a designer have to consider the yield, but he/she also has to make a trade-off between the yield and some other quality performance metrics of the design. For example, a designer often has to trade-off the yield against the power consumption. The multi-objective optimization problem targeted in this section is the trade-off between two performances with one of them being the yield. This problem appears widely in engineering design areas, since design parameters in practice often suffer from process and environmental variations. Not only in electrical engineering, such problem also widely appears in mechanical engineering [17], reliability engineering [18] and other disciplines [19, 20].

Mathematically, the problem leads to the following chance-constrained bi-objective stochastic optimization problem (CBSOP):

$$
\begin{aligned}
&\text{maximize } Y(x) = Pr(g_1(x, s) \geq 0, \ldots, g_k(x, s) \geq 0) \\
&\text{minimize } Q(x) = E_s[h(x, s)] \\
&\text{subject to } Y(x) \geq \theta, x \in [a, b]^d.
\end{aligned}
\tag{6.7}
$$

where

- x is the d-dimensional vector of design variables,
- s is a vector of random variables with zero mean values that models the manufacturing and environmental variations. If s is replaced by its mean value ($s = 0$), it implies that manufacturing and environmental variations are ignored,
- $g_i(x, s)$ ($i = 1, \ldots, k$) are k performance functions. A design meets the design requirements if $g_i(x, s) \geq 0$ for all i,
- $h(x, s)$ is the quality index function, which may be one of the $g_i(x, s)$ functions. For example, it may be the power consumption of an analog circuit,
- $Y(x)$ is the yield function,
- $Q(x)$ is the expected performance of the quality index,
- θ is the yield threshold.

Most existing approaches are in the framework of single-objective optimization, such as only optimizing $Y(x)$ (yield optimization), or optimizing $Q(x)$ with some constraint on $Y(x)$ (single-objective variation-aware design). That framework cannot provide trade-off information between $Y(x)$ and $Q(x)$. In industrial applications, the trade-off between $Y(x)$ and $Q(x)$ has received much attention and a Pareto front between $Y(x)$ and $Q(x)$ subject to a minimum value for $Y(x)$ needs to be obtained. Due to the stochastic value of s, evaluations of $Y(x)$ and $Q(x)$ require Monte-Carlo (MC) simulations, which makes the solution of Eq. (6.7) expensive. Currently, there are few existing methods which can solve the CBSOP in Eq. (6.7) in an effective and efficient manner.

In the following, a state-of-the-art OO-based MC method will be introduced to address this problem [21]. This is a general method focusing on efficient solution for expensive MC simulation-included optimization problems. Three main components are as follows:

(1) Latin Supercube sampling (LSS) [22], is applied, which is more efficient than LHS, and is more general and robust than MCMC methods [23] (for more details, see Chap. 5).
(2) The common practice of using the same number of samples for all the candidate solutions is not adopted when handling stochastic chance constraints. Instead, OO [2] is used to adaptively assign the number of samples to different candidate solutions.
(3) A two-phase optimization mechanism is developed. Although OO is very promising for efficiently selecting or ranking good candidates, its accuracy often cannot satisfy the requirement for objective optimization. The two-phase approach consists of a constraint satisfaction phase and an optimization phase. OO is applied in the constraint satisfaction phase using a small number of LSS samples. The purpose of this phase is to provide a well-distributed feasible initial population to the optimization phase. In the optimization phase, a large number of LSS samples are used to obtain more accurate function values. The MOEA/D [24] variant in [25] (MOEA/D-DE) is used as the optimization method (see Chap. 2).

The algorithm, multi-objective uncertain optimization with ordinal optimization, Latin Supercube sampling and parallel computation (MOOLP), will be introduced in the following.

6.6.1 The MOOLP Algorithm

6.6.1.1 Main Idea

To estimate $Y(x)$ and $Q(x)$ in (6.7), one has to do computer simulations which are often computationally expensive in practice. Thus, like in the previous sections, the goal is to reduce the number of simulations as much as possible. According to the pruning method introduced in Chap. 5 and multi-objective optimization, three considerations can be made:

- Constraints are more important than objectives. When a candidate solution does not meet the constraints, we do not deliver it as a final solution to the decision maker. Therefore, it is unnecessary to spend much effort to estimate the objective function values of an infeasible solution.
- In most engineering applications, if a solution does not meet the following deterministic constraints:

$$g_i(x, 0) \geq 0 \quad \text{for i=1, ..., k,} \tag{6.8}$$

that correspond to constraints without variations, then its $Y(x)$ will be zero or very low, and hence, it cannot meet the chance constraint:

$$Y(x) \geq \theta$$

for practical values of θ. Therefore, it is unnecessary to estimate $Y(x)$ when x is infeasible in terms of the deterministic constraints.

- It is relatively easy to find a set of (nearly) feasible solutions compared with the optimization problem itself. If most initial solutions are feasible and of good diversity in an EA, then its computational effort should be significantly reduced.

Based on the above considerations, the search can be divided into two phases: constraint satisfaction and objective optimization. The constraint satisfaction phase tries to find a set of nearly feasible solutions with a good diversity. The optimization phase takes the output of the constraint satisfaction phase as its initial population and searches for an approximate Pareto front (PF).

In the constraint satisfaction phase, it is not necessary to estimate $Q(x)$, and only estimate $Y(x)$ when necessary. More importantly, the goal is to provide an initial population to the optimization phase, so we only need to have a rough estimate of $Y(x)$, particularly for those solutions which are unlikely to be feasible since those solutions would be discarded later. Thus, it is very reasonable to use the OO method in this phase.

In the optimization phase, since its initial population is almost feasible, very sophisticated constraint handling techniques are not necessary. For simplicity, the penalty function approach can be used to maintain the feasibility. Because this phase produces the final solutions to the decision maker, a good number of simulations to each candidate solution are necessary to estimate its objective function values.

Since LSS is much more efficient than other sampling techniques, it can be used in both the constraint satisfaction and optimization phase. In addition, both, the evaluations of individuals in each iteration of evolutionary algorithms and their MC simulations are independent from each other, and, hence, parallel computation can be easily applied.

6.6.1.2 Phase 1: Constraint Satisfaction

The goal of the constraint satisfaction phase is to provide a good initial population to the optimization phase. We set the following two requirements for its final population Pop:

- For each solution x in Pop, $Y(x) \geq 0.9\theta$. The reason why we relax the chance constraint is that nearly feasible solutions can often provide very useful information for the optimization phase, and those solutions can be also helpful to increase the diversity of Pop.
- Pop should be of good diversity in the decision space. We use the following metrics to measure the uniformness of Pop:

$$Uni(Pop) = \sum_{x \in Pop} (\overline{dis} - dis(x, Pop\backslash\{x\}))^2 \qquad (6.9)$$

where $Pop\backslash\{x\}$ denotes the whole population Pop except the individual x and $dis(x, Pop\backslash\{x\})$ is the Euclidean distance between x and the individual of $Pop\backslash\{x\}$ closest to x, i.e.,

$$dis(x, Pop\backslash\{x\}) = min_{x' \in Pop\backslash\{x\}}||x - x'||,$$

and $\overline{dis}$ is the average of all these distances. The smaller the Uni value, the more uniform is the distribution of the points in Pop.

The constraint satisfaction starts with a population of N solutions randomly generated within $[a, b]^d$. At each generation, it produces one offspring solution for each solution in the current population by using the EA operators. In MOOLP, DE is used. To evaluate the qualities of these N new solutions, they can be split into two groups to make use of OO:

- Group A: it contains all the new solutions which do not meet the deterministic constraints $g_i(x, 0)$,
- Group B: it contains all the other new solutions.

For a solution x in Group A, we define its deterministic constraint violation $DCV(x)$ as the sum of the violations of each constraint. We do not need to do more simulations to estimate its Y value and simply set it to zero. We apply OO on Group B to estimate the Y values of its solutions. The main goal is to obtain a good ranking for the candidate solutions to make sure the selection is correct, while decreasing the necessary number of simulations as much as possible.

To update the population, each solution x and its offspring o will be compared. Solution o replaces x if, and only if, one of the following conditions is true:

- when x violates some deterministic constraints and $DCV(o) < DCV(x)$,
- when $DCV(x) = 0$, $Y(x) < 0.9\theta$ and $Y(x) < Y(o)$,

- when $Y(x) \geq 0.9\theta$, $Y(o) \geq 0.9\theta$, and Uni value of Pop will decrease if x is replaced by o.

Due to the nature of OO, for those solutions which severely violate the chance constraint, a very small number of simulations are allocated and, thus, their estimated Y values are not very accurate. However, it will not have a large effect since these solutions should be replaced during the constraint satisfaction, whereas the (nearly) feasible solutions for the initial population of the optimization phase have a good estimation accuracy because much more samples are allocated to them.

For simplicity, the current population Pop is used as the baseline when checking if the Uni value decreases by x is replaced by o. Standard DE is adopted in which all the replacements are conducted in parallel. Thus, a replacement based on the Uni change computed in our approach may not lead to the Uni decrease in some cases. However, since only a few solutions are replaced at each generation when the candidates move to the feasible space, this approach works well in practice. Normally, the Uni value of the current population increases first as the population moves towards the feasible region and then decreases as Uni becomes the major driving force. The constraint satisfaction phase stops when the Uni value is smaller than that of the initial population or the number of iterations used exceeds a predefined number.

The constraint satisfaction phase is summarized as follows:

Step 1: Randomly sample N candidate solutions uniformly distributed in $[a, b]^d$ to form the initial population Pop. Compute the DCV values of all the solutions, divide them into two groups and set/estimate their Y values accordingly. If no solution in Pop meets all the deterministic constraints, the best solution is that with the smallest DCV value. Otherwise, the best solution is that with the largest Y value.

Step 2: Use the DE operators to create an offspring for each solution in Pop. Compute the DCV values of all the offsprings, divide them into two groups and set/estimate their Y values accordingly.

Step 3: Compare each solution with its offspring, and update Pop.

Step 4: If the stopping condition is met, output Pop. Otherwise, go to Step 2.

6.6.1.3 Phase 2: Objective Optimization

MOEA/D is used for multi-objective optimization in the optimization phase. A general framework of MOEA/D is proposed in [24] and has been introduced in Chap. 2. In MOOLP, the Tchebycheff approach and a penalty function technique are used for transforming (6.7) into N unconstrained single objective optimization problems. More specifically, the kth subproblem is formulated as the minimization of:

$$F^k(x) = max \left\{ \frac{k}{N-1}|Y(x) - Y^*|, \frac{N-k-1}{N-1}|Q(x) - Q^*| \right\}$$
$$+ \rho \times (max\{\theta - Y(x), 0\}) \tag{6.10}$$

where Q^* and Y^* are the largest values of $Q(x)$ and $Y(x)$ in the feasible region.

It is worth noting that as Q^* and Y^* are usually unknown before the search, the algorithm replaces them by the largest Q and Y values found during the search. The same number of LSS samples is used for each candidate solution to estimate its Y and Q values. The number of the LSS samples should be large enough to obtain a good accuracy.

In MOOLP, MOEA/D-DE [25] is adopted. During the search, MOEA/D-DE maintains:

- a population of N solutions $x^1, \ldots, x^N \in [a, b]^d$, where x^i is the current solution of the ith subproblem; and their estimated Q and Y values,
- Q^* and Y^*, i.e., the largest Q and Y values found so far, respectively.

Let $B(i)$, $i \in \{1, \ldots N\}$ contain the indices of the T closest neighbours of x^i, defined by the Euclidean distances between the ith weight vectors and the others. The kth subproblem is a neighbour of the ith problem if $k \in B(i)$. Neighbouring problems have similar objective functions, so their optimal solutions should be similar too. Before applying MOEA/D, we should set the value of the neighbourhood size T. We also need to set the value of n_r, the maximal number of solutions allowed to be replaced by a single new solution.

The algorithm is summarized as follows:

1. **Initialization:**

    1.1: Set the output of Phase 1 as the initial population. Use n_{ac} LSS samples to estimate the Y and Q values of each individual.
    1.2: Randomly assign the solutions in the initial population to the subproblems.
    1.3: Initialize Q^* and Y^* as the largest Q and Y function values, respectively, in the initial population.

2. **Update:**
    For $i = 1, \ldots, N$

    2.1 **Selection of the mating pool:**
    Generate a uniformly distributed random number $rand$ in (0,1]. Set

    $$P_{nb} = \begin{cases} B(i), & \text{if } rand < \delta, \\ \{1, \ldots, N\}, & \text{otherwise.} \end{cases} \quad (6.11)$$

    2.2 **Reproduction:**
    Set $x^{best} = x^i$; randomly select two indexes, r_1 and r_2, from P_{nb}; and generate a new solution $\overline{y}$ using the DE operators. Then, perform a polynomial mutation [26] on $\overline{y}$ with probability p_m to produce a new solution y (see Chap. 2 for these operators).
    2.3 Use LSS sampling to estimate $Q(y)$ and $Y(y)$.
    2.4 **Update Q^* and Y^*:**
    If $Q(y) > Q^*$, set $Q^* = Q(y)$. If $Y(y) > Y^*$, set $Y^* = Y(y)$.

2.5 **Update of Solutions:**
Set $c = 0$ and then do the following:

(1) If $c = n_r$ or P_{nb} is empty, go to Step 3. Otherwise, randomly pick an index j from P_{nb}.
(2) If $F^i(y) < F^i(x^j)$, then set $x^j = y$, and $c = c + 1$.
(3) Remove j from P_{nb} and go to (1).

End

3. **Stopping**
If the stopping criteria (e.g., a certain number of iterations) are satisfied, then stop and output $\{x^1, \ldots, x^N\}$, $\{Y(x^1), \ldots, Y(x^N)\}$ and $\{Q(x^1), \ldots, Q(x^N)\}$. Otherwise go to Step 2.

6.6.2 Experimental Results

In this subsection, MOOLP is used for the analog circuit shown in Fig. 6.7. The circuit is designed in a 90 nm CMOS process with 1.2 V power supply. The specifications are:

$$
\begin{aligned}
&\text{DC gain} \geq 60\,\text{dB} \\
&\text{gain-bandwidth product} \geq 45\,\text{MHz} \\
&\text{phase margin} \geq 60° \\
&\text{output swing} \geq 2\,\text{V} \\
&\text{power} \leq 2.5\,\text{mW} \\
&\text{area} \leq 180\,\mu\text{m}^2
\end{aligned}
\tag{6.12}
$$

$Y(x)$ is the probability that a design meets all the above specifications. The chance constraint is $Y(x) \geq 85\,\%$. The goal is to maximize $Y(x)$ and minimize the power consumption ($Q(x)$). There exist 21 design variables, including transistor widths W_i, transistor lengths L_i, the compensation capacitance Cc and the biasing current I_b. The bounds of the design variables are:

$$
\begin{aligned}
&0.12\,\mu\text{m} \leq W_i \leq 800\,\mu\text{m} \\
&0.1\,\mu\text{m} \leq L_i \leq 20\,\mu\text{m} \\
&0.1\,\text{pF} \leq Cc \leq 50\,\text{pF} \\
&0.05\,\text{mA} \leq I_b \leq 50\,\text{mA}
\end{aligned}
\tag{6.13}
$$

All transistors must be in the saturation region. The number of process variation variables is 52, which are all normally distributed. To estimate $Y(x)$ and $Q(x)$, all these 52 variation variables need to be considered. The experiments were implemented in MATLAB running on a PC with an 8-core CPU, 4GB RAM and the Linux operating system. The electrical simulator HSPICE is used as the circuit performance evaluator. Details of the parameter settings are in [21]. The wall-clock time spent in this problem is 112 h. The post-processed PF is shown in Fig. 6.9.

For comparison, MOEA/D with PMC simulation and only one optimization phase (MOEA/D with penalty function to handle the chance constraint) for solving this problem is also used. All the parameter settings are exactly the same as those in the MOOLP approach. This method could not produce any feasible solutions after 112 h of wall-clock time if just a penalty function for the chance constraint $Y(x) \geq 85\%$ is used. This result shows that the two-phase approach is necessary for this real-world problem. Then, $g_i(x, 0) \geq 0$ is included (i.e., the specifications without considering process variations) in the constraints and add them together with the chance constraint to the penalty function. After 112 h, the solutions generated and post-processed like in MOOLP are also shown in Fig. 6.9. It is clear that the approximate PF by the traditional method is much worse than that generated by MOOLP:

- the largest obtained yield is lower than that obtained by MOOLP,
- there are many gaps in the PF compared with that obtained by MOOLP,
- all points in the PF are dominated by those provided by MOOLP.

These performance differences are due to the following reasons:

- without the constraint satisfaction phase, many samples are wasted in infeasible solutions, and less computational effort is allocated to important solutions.
- LSS can provide better samples than PMC, which makes MOOLP estimate the Y and Q values more accurately.

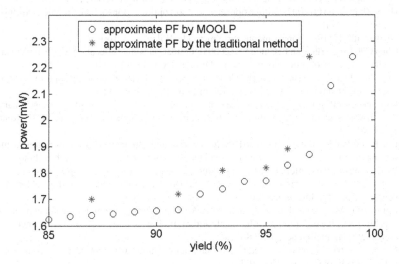

Fig. 6.9 Comparison of the power-yield trade-off of the amplifier by MOOLP and the traditional method

6.7 Summary

This chapter has introduced the ordinal optimization-based Monte-Carlo method for yield optimization, and single and multi-objective variation-aware analog circuit sizing methods. These problems are common in real-world applications, especially in the manufacturing and reliability engineering fields. The key techniques are OO and efficient evolutionary search mechanisms. OO aims to intelligently assign the computational budgets in order to correctly select promising candidates from the population with as few MC simulations as possible. Efficient evolutionary search techniques decrease the number of necessary iterations by enhancing the exploitation ability of a standard EA. Their cooperation has been shown by the ORDE method for the targeted problem. Multi-objective variation-aware design is an important problem but is quite difficult. A state-of-the-art method based on OO has been provided, including a new search mechanism which can make full use of OO and the LSS method.

References

1. McConaghy T, Palmers P, Gao P, Steyaert M, Gielen G (2009a) Variation-aware analog structural synthesis: a computational intelligence approach. Springer Verlag, New York
2. Ho Y, Zhao Q, Jia Q (2007) Ordinal optimization: soft optimization for hard problems. Springer-Verlag, New York
3. Niederreiter H (1992) Quasi-monte carlo methods. Wiley, New York
4. Chen CH, Lin J, Yücesan E, Chick SE (2000a) Simulation budget allocation for further enhancing the efficiency of ordinal optimization. Discrete Event Dyn Syst 10(3):251–270
5. Chen H, Chen C, Yucesan E (2000b) Computing efforts allocation for ordinal optimization and discrete event simulation. IEEE Trans Autom Control 45(5):960–964
6. Moscato P (1989) On evolution, search, optimization, genetic algorithms and martial arts: Towards memetic algorithms. Caltech concurrent computation program, C3P. Report 826:1989–2057
7. Liu B, Fernández F, Gielen G (2010) An accurate and efficient yield optimization method for analog circuits based on computing budget allocation and memetic search technique. In: Proceedings of the conference on design, automation and test in Europe, pp 1106–1111
8. Lagarias J, Reeds J, Wright M, Wright P (1998) Convergence properties of the Nelder-Mead simplex method in low dimensions. SIAM J Optim 9:112–147
9. Epitropakis M, Plagianakos V, Vrahatis M (2008) Balancing the exploration and exploitation capabilities of the differential evolution algorithm. In: Proceedings of IEEE world congress on computational intelligence, pp 2686–2693
10. Noman N, Iba H (2008) Accelerating differential evolution using an adaptive local search. IEEE Trans Evol Comput 12(1):107–125
11. Das S, Abraham A, Chakraborty U, Konar A (2009) Differential evolution using a neighborhood-based mutation operator. IEEE Trans Evol Comput 13(3):526–553
12. Liu B, Fernández F, Gielen G (2011a) Efficient and accurate statistical analog yield optimization and variation-aware circuit sizing based on computational intelligence techniques. IEEE Trans Comput Aided Des Integr Circuits Syst 30(6):793–805
13. Lampinen J, Zelinka I (2000) On stagnation of the differential evolution algorithm. In: Proceedings of 6th international mendel conference on, soft computing, pp 76–83

14. Deb K (2000) An efficient constraint handling method for genetic algorithms. Comput Methods Appl Mech Eng 186(2):311–338
15. Graeb H (2007) Analog design centering and sizing. Springer Publishing Company, Incorporated, Dortrecht, Netherlands
16. Swidzinski J, Chang K (2000) Nonlinear statistical modeling and yield estimation technique for use in Monte Carlo simulations [microwave devices and ICs]. IEEE Trans Microw Theory Tech 48(12):2316–2324
17. Mercado LL, Kuo SM, Lee TY, Lee R (2005) Analysis of RF MEMS switch packaging process for yield improvement. IEEE Trans Adv Packag 28(1):134–141
18. Chan H, Englert P (2001) Accelerated stress testing handbook. IEEE Press, New York
19. Poojari C, Varghese B (2008) Genetic algorithm based technique for solving chance constrained problems. Eur J Oper Res 185(3):1128–1154
20. Liu B (2002) Theory and practice of uncertain programming. PhysicaVerlag, Berlin
21. Liu B, Zhang Q, Fernández F, Gielen G (2013a) An efficient evolutionary algorithm for chance-constrained bi-objective stochastic optimization and its application to manufacturing engineering. IEEE Trans Evol Comput(To be published) doi:10.1109/TEVC.2013.2244898.
22. Owen A (1998) Latin supercube sampling for very high-dimensional simulations. ACM Trans Model Comput Simul (TOMACS) 8(1):71–102
23. Singhee A, Rutenbar R (2009) Novel algorithms for fast statistical analysis of scaled circuits. Springer Verlag, New York
24. Zhang Q, Li H (2007) MOEA/D: a multiobjective evolutionary algorithm based on decomposition. IEEE Trans Evol Comput 11(6):712–731
25. Li H, Zhang Q (2009) Multiobjective optimization problems with complicated Pareto sets, MOEA/D and NSGA-II. IEEE Trans Evol Comput 13(2):284–302
26. Deb K, Pratap A, Agarwal S, Meyarivan T (2002) A fast and elitist multiobjective genetic algorithm: NSGA-II. IEEE Trans Evol Comput 6(2):182–197

Chapter 7
Electromagnetic Design Automation: Surrogate Model Assisted Evolutionary Algorithm

This chapter starts the third part of the book, simulation-based electromagnetic design automation, including high-frequency integrated passive component synthesis, mm-wave linear amplifier synthesis, mm-wave nonlinear amplifier synthesis and complex antenna synthesis. From the CI point of view, this part concentrates on surrogate model assisted evolutionary algorithm (SAEA). SAEA is a cutting-edge research topic in the CI field. Before introducing SAEA, almost no general and practical design automation method has been published for the above topics in the EDA field, yet it is very much needed in industry.

In recent years, the demand for high-data-rate wireless communication systems is constantly increasing. Low-GHz RF ICs (i.e., below 5 GHz) are not able to support these high-data-rate communications [1]. Hence, the design and optimization methods for mm-wave RF ICs are attracting a lot of attention recently. In particular, research on RF building blocks for 40–120 GHz and beyond is increasing drastically. For example, 60 GHz RF ICs are widely used for uncompressed HDTV and 94 GHz RF ICs are used for mm-wave imaging systems. Antenna, as a special passive component, is as important as integrated circuits in modern communication systems. Chapters 7–10 provide the first efficient single-objective global optimization and multi-objective optimization techniques for these problems. This part also elaborates the methodology to construct an SAEA, including fundamentals and new advances.

This chapter defines the problems, reviews the current research of the targeted problems and the challenges encountered, and introduces the computational intelligence background behind the synthesis problems. It is organized as follows. Section 7.1 provides a global picture of simulation-based electromagnetic design automation, which makes the main challenges clear. From Sects. 7.2 to 7.3, the related current research is reviewed and the challenges for integrated passive component synthesis, mm-wave integrated circuit synthesis and antenna synthesis are elaborated. The key method to address the challenges in electromagnetic design automation is SAEA, which adaptively integrates machine learning (supervised learning) into evolutionary algorithms. Section 7.4 reviews the benefits and challenges of surrogate

B. Liu et al., *Automated Design of Analog and High-frequency Circuits*,
Studies in Computational Intelligence 501, DOI: 10.1007/978-3-642-39162-0_7,
© Springer-Verlag Berlin Heidelberg 2014

Table 7.1 UHF, SHF and EHF in microwaves

Frequency	Wavelength	Designation
300 MHz–3 GHz	10 cm–1 m	Ultra high frequency (UHF)
3 GHz–30 GHz	1–10 cm	Super high frequency (SHF)
30 GHz–300 GHz	1 mm–1 cm	Extremely high frequency (EHF)

model assisted evolutionary algorithms. Section 7.5 introduces the main learning machine used, Gaussian Process. Section 7.6 briefly introduces artificial neural network. This chapter is summarized in Sect. 7.7.

7.1 Introduction to Simulation-Based Electromagnetic Design Automation

Electromagnetic design automation concentrates on optimization of structures (e.g., integrated transformers, ICs, antennas), where the computationally expensive electromagnetic simulation is needed. We can generally call them microwave structures. Some frequency bands are shown in Table 7.1 Because in EHF, the wavelength is in the order of millimeter, it is also called mm-wave frequency. In this frequency range, the wavelength is often comparable to the size of the microwave structures and electromagnetic analysis is a must.

Simulation-based electromagnetic design automation problems are global optimization problems. Let us consider as an example of single-objective constrained optimization problem, the synthesis of a 60 GHz two-stage power amplifier with the following specifications:

$$
\begin{aligned}
\text{maximize} \quad & \text{1 dB compression point} \\
\text{s.t.} \quad & \text{power gain} \geq 15 \, \text{dB} \\
& \text{power added efficiency} \geq 15 \, \%
\end{aligned}
\tag{7.1}
$$

Since evolutionary computation is independent of the practical applications, a natural question is why the methods developed in Chaps. 2–6 cannot be used for these problems. The answer is that the required electromagnetic (EM) simulation used to calculate the performance values to evaluate the objective functions and constraints, is computationally expensive. Simulation-based optimization is black-box optimization. In recent years, evolutionary computation (EC) is becoming a standard approach for complex black-box optimization due to its high ability to converge close to the globally optimal solution, its suitability for multimodal problems and its good robustness. In many cases, evolutionary algorithms (EAs) are the only feasible methods for simulation-based optimization. However, EA often requires hundreds to thousands of candidate solution evaluations in its optimization process, many more than those required by deterministic methods for local optimization. If EAs, such as the methods from Chaps. 2 to 6 are applied, high-quality designs can indeed be obtained, but the time consumed may be intractable, which is not practical for industrial applications. For example, when differential evolution is applied, the synthesis of a linear amplifier at 100 GHz takes nearly 10 days (see Chap. 9). Although with high-quality results, this method cannot be used in industry because it is too slow. On the other hand, traditional mathematical programming methods require relatively few candidate evaluations. The fatal problem is that most of them can only achieve local optimization, and the optimization quality highly depends on the starting point, which should be provided by the designers. Without a good starting point, the quality of the result can be bad, because the component, circuits and antenna synthesis problems are often multimodal. Hence, a high design quality cannot be satisfied when traditional deterministic optimization methods are selected.

Therefore, it is clear that the goal of the methods developed for EM simulation-based design automation is to obtain high-quality designs which are comparable to those obtained by plain EM simulations in EA loops, but computationally much cheaper and, therefore, to perform the synthesis in a practical time. In the following sections, we will review existing approaches and investigate the bottlenecks of electromagnetic design automation.

7.2 Review of the Traditional Methods

7.2.1 Integrated Passive Component Synthesis

Integrated passive component synthesis methodologies, which serve as the foundation of RF and mm-wave integrated circuit synthesis, are reviewed first. It is well known that on-chip passive components, e.g., inductors and transformers, strongly influence the circuit performances of RF ICs [2]. For example, the loss of a transformer has a large impact on the power-added efficiency and the output power of a power amplifier (PA). Therefore, the synthesis of passive components, including both the sizing and the layout optimization, is a critical problem in RF and mm-wave IC design automation.

Most of the traditional computer-aided design optimization methodologies for integrated passive components can be classified into four categories: (1) equivalent circuit model and global optimization-based (ECGO) methods [3, 4]; (2) EM-simulation and global optimization-based (EMGO) methods [2]; (3) off-line surrogate model, EM-simulation and global optimization-based (SEMGO) methods [5] and (4) surrogate model and local optimization-based (SMLO) methods [6–10]. These will now be described in more detail.

- The ECGO methods [3, 4] depend on an equivalent circuit model of the passive component to obtain the performances of the microwave structure. Their advantage is high efficiency. The synthesis of a 5 GHz inductor considering process variations, which requires many performance evaluations, has been achieved successfully and efficiently by ECGO [3]. On the other hand, when the frequency is high, equivalent circuit models available in the microwave area are typically not accurate enough or difficult to find. Hence, even with global optimization algorithms, the synthesis of high-frequency components may also fail as the equivalent circuit models used may not reflect well the performances of the microwave structures.

- The EMGO methods [2] can provide an accurate performance analysis of the RF structure because they use EM simulations. Combined with global optimization algorithms, the quality of the solution is often the best among all the available methods, especially in high-frequency RF component synthesis. However, its major bottleneck is the high computational cost of the EM simulations, limiting their use in practice [6].

- Reference [5] represents a surrogate-model EMGO (SEMGO), which is an important progress compared to EMGO. Surrogate modeling is an engineering method used when an outcome of interest cannot be directly measured easily (in this case, the simulation is too time consuming), so an approximate model of the outcome is used instead (in this case, it is computationally cheap). SEMGO uses an off-line artificial neural network (ANN) model to enhance the speed of the standard EMGO. In [5] the surrogate model is first trained to approximate the performance of the RF structure before optimization. Then, the optimization algorithm uses this surrogate model as the performance evaluator to find the optimal design. The training data are generated uniformly in the design space and the corresponding performances are obtained by EM simulations. When combined with global optimization algorithms, this method has the ability of global search. However, the training data generation process in this method is expensive and the constructed ANN model may not always be reliable (see Chap. 8).

- The SMLO methods [6–10] combine the efficiency of ECGO with the accuracy of the EM simulations from EMGO. Figure 7.1 shows the general flow. First, a coarse model, either an equivalent circuit model or a coarse-mesh EM simulation, is constructed and optimized. Then, some base vectors in the vicinity of the optimal point of the coarse model are selected as the base points to train a more detailed surrogate model, whose purpose is to predict the performances of the microwave structure. Finally, the surrogate model is used to optimize the RF component, whose result is verified by the fine model using expensive high-fidelity EM simulations. The data received from the fine simulations will update the surrogate model to make

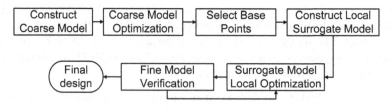

Fig. 7.1 Flow of the SMLO methods

it more accurate. In the development of the SMLO methods, some works have been presented focusing on selecting the coarse model [6, 8] and the surrogate model [6, 9].

SMLO, however, highly depends on the accuracy of the coarse model, which leads to two significant challenges for high-frequency RF passive component synthesis. First, the optimal solution of the coarse model defines the search space and the constructed surrogate model is only accurate in that space, because the base points are selected around it [6]. The success of SMLO comes from the basic assumption that the optimal point of the coarse and fine models are not far away in the design space, as shown in [6–10]. However, this assumption only holds when the coarse model is accurate enough. Although it has been shown that SMLO can solve RF component synthesis well at comparatively low frequencies [6–10] (e.g., 10 GHz), for passive components in high-frequency RF ICs (e.g., 60 GHz), this assumption is often not true. In many cases, it is difficult to find an accurate equivalent circuit model or to decide the mesh density of the EM simulation to trade the accuracy and the efficiency off. The second challenge is that SMLO can only do local search, which is not suited for synthesis with strong requirements. This is not only because of the fact that the current SMLO methods use local optimization algorithms, but also because of the fact that the search space is decided first by the coarse model [6–10]. Therefore, using global optimization algorithms makes little sense.

In summary, ECGO and SMLO work well in comparatively low-frequency RF component synthesis, but their high dependence on the accuracy of the equivalent circuit or coarse model limits their use for the synthesis of high-frequency mm-wave structures. EMGO can provide high-quality results even when the frequency is high, but it is too CPU time intensive. Although SEMGO [5] makes a good progress beyond EMGO, to the best of our knowledge, its speed enhancement and robustness still need to be improved.

7.2.2 RF Integrated Circuit Synthesis

Traditional RF IC design automation methods focus on low-GHz synthesis [11–19] by employing ECGO in Sect. 7.2.1. The architecture of most of these methods is shown in Fig. 7.2. Compared with the low-frequency analog circuit sizing flow, a key part is the generation of the parasitic-aware model of the passive components. In RF IC designs at low-GHz frequencies, a simple lumped model is often extracted to mimic the behavior of the key passive components (transformer, inductor). Regression methods are then used to fit the EM simulation results (S-parameters) to parasitic-included equivalent circuit models. The generated passive component models are accurate at low-GHz frequencies and computationally efficient.

To make the parasitic-aware model reliable in providing the correct performances for different design parameters, a strictly enforced layout template is often necessary. Some works use the parasitic corner, rather than a strict layout template, to

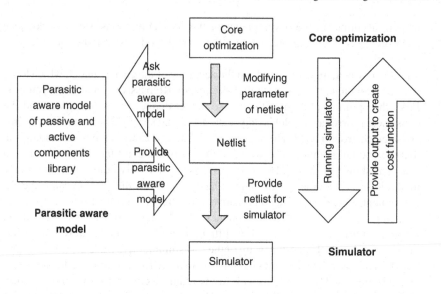

Fig. 7.2 Framework of parasitic-aware optimization to RF IC (from [12])

improve the flexibility of the generated layout for circuits below 10 GHz. In the development of the optimization kernel, evolutionary algorithms (EAs) are introduced in RF IC synthesis to achieve global search, getting very good results [15, 16]. Park (2003) uses Particle swarm optimization (PSO) and [18] introduces the non-dominated genetic algorithm (NSGA) to RF IC synthesis in order to achieve multi-objective optimization.

7.2.3 Antenna Synthesis

Antenna synthesis through global optimization has been widely applied. Evolutionary computation-based synthesis methods are well investigated [20–22]. In electromagnetism, optimization problems generally require high computational resources and involve a large number of unknowns [21]. They are usually characterized by non-convex functions and a continuous space for strategies based on evolutionary algorithms [21]. Due to the wide application of EC in antenna design, antenna benchmark problems for testing newly proposed EAs have been proposed [23]. DE and PSO are the most popular algorithms in the antenna optimization area [22, 23]. Besides using DE and PSO, various improvements based on standard DE and PSO have been investigated and compared with each other [24–26]. Goudos (2011) proposes a self-adaptive DE, and shows the advantages of DE compared with PSO in general. From the computational intelligence point of view, PSO can be recognized as a kind of DE, and their basic ideas have many similarities. In 2011, [27] applied a covariance matrix adaptation evolution strategy (CMA-ES) and showed a robust and more efficient alternative to PSO for antenna synthesis.

With this background, two research directions have attracted much attention: (1) improving DE and PSO for the specific use of antenna global optimization [25, 26], whose goal is to achieve better antenna designs; (2) extending constraint satisfaction and single-objective constrained optimization to multi-objective optimization [28, 29], since, like in integrated circuits, the interesting antenna specifications have trade-off relations between each other (e.g., gain and area consumption).

7.3 Challenges of Electromagnetic Design Automation

According to Sect. 7.1, there are at least two key elements for a good electromagnetic design automation method, which are:

(1) effective global optimization capacity,
(2) effective speed enhancement.

The former has been intensively investigated, and the latter is the main bottleneck which prevents many available methods coming to practice.

For integrated passive components, a speed enhancement approach, commonly employed by designers, is to use the lumped equivalent circuit models to replace the computationally expensive EM simulations (ECGO and SMLO). Other methods, EMGO and SEMGO, suffer from too long computational time, especially when the number of design variables is large.

Parasitic-aware lumped equivalent circuit models for passive components that accurately match the EM simulation results are often difficult to find at frequencies between say 40 GHz and above 100 GHz due to the distributed effects. When the wavelength is comparable to the size of the devices, using a lumped model is often not accurate any more. Hence, when lumped equivalent circuit models are employed, available RF integrated circuit design automation methods are limited to low-GHz applications. Because the speed enhancement method for RF IC synthesis cannot be extended to mm-wave integrated circuit synthesis, and because directly including the EM simulations in each performance evaluation is too CPU time expensive, no efficient method for mm-wave integrated circuit synthesis existed before SAEA was introduced. The only way left to mm-wave circuit designers is the "experience and simulation verification" method, which is at odds with today's high-performance and tightening time-to-market requirements.

For antenna design, no matter in which frequency, equivalent circuit models are only available for very simple antennas. That is also a reason to explain why EC-based methods are quite developed for antenna synthesis. It is clear that developing optimization frameworks with higher search ability is very important. However, efficiency enhancement has not been well investigated yet but is very important in antenna synthesis, since it touches the fundamental issues. Most of the current antenna synthesis methods grant a long computational time because there is almost no other design method to achieve a very high quality design. EM simulation is often a must in antenna synthesis, and the time consumption to perform EM simulation for a candidate solution varies from a few seconds to tens of minutes [30], or even several hours. The EC-based synthesizer often needs 30,000–20,0000 evaluations (EM simulations) for antenna problems [30]. It is clear that drastically enhancing the efficiency without or slightly sacrificing the performance is extremely important for antenna synthesis. John (2009) uses the ParEGO method [31] from the computational intelligence field for the multi-objective optimization of antennas. Siah (2004) used the EGO method [33] in computational intelligence for single-objective antenna synthesis. In these methods, surrogate modeling and Gaussian Process machine learning are introduced to antenna optimization. They contribute to the efficiency enhancement of antenna synthesis, which is one step further for the EC-based intelligent algorithms needed for industrial antenna design. However, the dimensionality that can be handled by the methods in [30, 32] is not high. The examples in [30, 32] have 3–5 design variables. However, for practical antenna design, the design parameters can vary from a few to more than 30. Therefore, the efficient global optimization methods in [30, 32] can hardly be used for complex antenna design. This is also a problem for mm-wave integrated circuit design, which also has tens of design parameters. In the CI field, efficient global optimization method for medium-scale (20–50 dimensions) computationally expensive optimization problems is also an open area. This is called the "curse of dimensionality" in the field. We will present solutions for this in Chaps. 9 and 10.

To sum up, for high-frequency integrated passive component synthesis, we need efficient global optimization methods for small-scale computationally expensive problems. For mm-wave integrated circuits and complex antennas, we need efficient global optimization methods for medium-scale computationally expensive problems.

7.4 Surrogate Model Assisted Evolutionary Algorithms

As has been discussed, the requirements and also challenges encountered in EM simulation-based design automation can be summarized as: (1) good global optimization ability, (2) high efficiency, (3) scalable to medium- and high-dimensional problems. In the computational intelligence field, such problems belong to the research area of computationally expensive global optimization, which is a cutting-edge research topic. The surrogate model assisted evolutionary algorithm (SAEA) is the computational intelligence background behind EM simulation-based design automation. The goal of SAEA is to obtain the near globally optimal solution with a reduced number of expensive function evaluations [34], which addresses the first two requirements. The third requirement, scalability to medium-dimensional problems, is still an open question even in today's SAEA research. Note that in SAEA research, optimization problems with less than 10 decision variables are often considered as small-scale problems, while problems with 20–50 decision variables are typically medium-scale problems, and problems with larger than 100 decision variables are large-scale problems. Hence, integrated passive component synthesis typically belongs to small-scale problems, while mm-wave integrated circuit synthesis and complex antenna synthesis are often medium-scale problems. In the following, a general review of SAEA research will be presented.

SAEA is a promising approach for dealing with expensive optimization problems. These algorithms employ surrogate models to replace the computationally expensive function evaluations. Since the construction of the surrogate model and its use to predict the function values cost much less effort than embedding the expensive function evaluator within the optimizer, the computational cost can be significantly reduced. Note that the surrogate modeling and the evolutionary algorithms must work in a cooperative manner in SAEA. For example, the SEMGO and the SMLO method also include surrogate models, but they are not SAEA. The reason is that in these two methods, a surrogate model which guides the evolutionary search, especially global search, is built in a preprocessing phase (i.e., off-line surrogate model), and if there is on-line updating, it only improves the local accuracy of the model.

In recent years, many surrogate model construction methods and the corresponding SAEAs were investigated. Among them, Gaussian Process (GP) or Kriging, response surface methods (RSM), artificial neural networks (ANN), support vector machines (SVM), radial basis function (RBF) models have been widely used [34–37]. More recently, some schemes in surrogate model assisted single-objective optimization were extended to EAs for multi-objective expensive optimization problems [31, 35, 38, 39]. Early methods, such as some examples in [34], do not consider the model uncertainty in the evolution process. With a number of samples with exact function evaluations serving as the training data, a surrogate model is constructed and the optimal candidate solutions based on surrogate model prediction are evaluated by exact function evaluators. The surrogate model is then updated and the above step is repeated until convergence. To address the issue of incorrect convergence of the above method, individual and generation-based control are introduced to consider the estimation error [40]. The goal is to determine when the surrogate model must be updated to avoid false optima. These are the prediction-based methods.

Besides prediction-based methods, prescreening has also been introduced into SAEA (e.g., expected improvement, probability of improvement) [33, 41]. Instead of expecting the surrogate model prediction to replace the exact function evaluation (the model uncertainty should be as small as possible to this end), prescreening methods aim to select the possible promising candidates from the newly generated candidate solutions utilizing the prediction uncertainty. Because both the EA and the prescreening methods contribute to the global search, prescreening-based methods can often detect the globally optimal or near optimal solutions efficiently for small-scale problems. Note that most prescreening methods, such as the expected improvement, are specially defined for the GP model, because they rely on the Gaussian distribution of the prediction. Successful prescreening-based SAEA examples are reported in [35, 42]. These methods can often obtain high-quality solutions in a relatively small number of function evaluations. However, large dimensionality is a challenge for prescreening-based methods. For medium- or large-scale problems, the surrogate model may fail to give predictions of acceptable accuracy within an affordable number of sample points over a high-dimensional search space [35]. Prescreening methods then cannot work as well as they do for small-scale problems.

As said above, both small-scale (<10) and medium-scale (with 20–50 decision variables) expensive optimization problems have to be addressed in EM simulation-based design automation. Recently, research on medium-scale SAEAs is attracting more attention [35, 43–45]. Reference [43] uses the prediction-based method. A reasonably good result is often obtained, but it costs 8,000 exact function evaluations for typical selected benchmark problems in the EC field. Reference [44] uses combined prediction and prescreening-based methods, and its solution quality and efficiency are comparable to those in [43]. These methods use local surrogate models and memetic algorithms. In particular, after using the evolutionary operators to locate the new candidate solutions in a global manner, a local search phase is included to refine these candidate solutions based on cheap local surrogate models. Exact function evaluations are then performed on the selected promising candidates and the surrogate models are updated. Reference [35] uses the prescreening-based method and the computational load is set to 1,000 exact function evaluations. Promising results have been obtained for some benchmark problems, but for others the results still need to be improved [35], especially for multimodal problems. Reference [35] also shows that the four commonly used prescreening methods have advantages on different problem landscapes, and none of them work well for all problems. Reference [45] invents a new surrogate model-aware evolutionary search mechanism, which considers both good optimization ability and good surrogate model construction. It achieves 2–8 times speed enhancement compared to [35, 43, 44] for 20 and 30-dimensional problems. A dimension reduction technique, Sammon mapping, has also been introduced into SAEA for medium-scale expensive optimization problems. However, the solution quality for 50-dimensional problems still needs to be improved. To summarize, much effort is required to develop both effective and efficient SAEAs for medium-scale and large-scale computationally expensive optimization problems.

In EM simulation-based design automation, the synthesis of integrated passive components and some antennas are small-scale expensive optimization problems. The synthesis of mm-wave linear amplifiers is a medium-scale expensive optimization problem but with decomposable structure, which means that it is possible to transform it to small-scale but more complex expensive optimization problems. The synthesis of non-linear mm-wave circuits and complex antennas are medium-scale expensive optimization problem and without decomposable structure. They will be addressed in Chaps. 8–10, respectively.

7.5 Gaussian Process Machine Learning

Gaussian Process machine learning is the main method used to build surrogate models in this book, due to its solid mathematical foundation. Typically, classification and prediction are the two main applications areas of machine learning methods. The latter is used in EM simulation-based design automation. In this section, the basics of Gaussian Process (GP) modeling and prediction are introduced.

7.5.1 Gaussian Process Modeling

To model an unknown function $y = f(x)$, $x \in R^d$, the GP modeling assumes that $f(x)$ at any point x is a Gaussian random variable $N(\mu, \sigma^2)$, where μ and σ are two constants. For any x, $f(x)$ is a sample of $\mu + \varepsilon(x)$, where $\varepsilon(x) \sim N(0, \sigma^2)$. For any $x, x' \in R^d$, the correlation between $\varepsilon(x)$ and $\varepsilon(x')$, $c(x, x')$, depends on $x - x'$. More precisely,

$$c(x, x') = \exp\left(-\sum_{i=1}^{d} \theta_i |x_i - x_i'|^{p_i}\right), \tag{7.2}$$

where parameter $1 \leq p_i \leq 2$ is related to the smoothness of $f(x)$ with respect to x_i, and parameter $\theta_i > 0$ indicates the importance of x_i on $f(x)$. More details about GP modeling can be found in [46].

7.5.1.1 Hyper Parameter Estimation

Given K points $x^1, \ldots, x^K \in R^d$ and their f-function values $y^1, \ldots, y^K$, then the hyper parameters $\mu, \sigma, \theta_1, \ldots, \theta_d$, and $p_1, \ldots, p_d$ can be estimated by maximizing the likelihood that $f(x) = y^i$ at $x = x^i$ ($i = 1, \ldots, K$) [33]:

$$\frac{1}{(2\pi\sigma^2)^{K/2}\sqrt{det(C)}} \exp\left[-\frac{(y - \mu\mathbf{1})^T C^{-1}(y - \mu\mathbf{1})}{2\sigma^2}\right] \tag{7.3}$$

where C is a $K \times K$ matrix whose (i, j)-element is $c(x^i, x^j)$, $y = (y^1, \ldots, y^K)^T$ and $\mathbf{1}$ is a K-dimensional column vector of ones.

To maximize (7.3), the values of μ and σ^2 must be:

$$\hat{\mu} = \frac{\mathbf{1}^T C^{-1} y}{\mathbf{1}^T C^{-1} \mathbf{1}} \tag{7.4}$$

and

$$\hat{\sigma}^2 = \frac{(y - \mathbf{1}\hat{\mu})^T C^{-1}(y - \mathbf{1}\hat{\mu})}{K}. \tag{7.5}$$

Substituting (7.4) and (7.5) into (7.3) eliminates the unknown parameters μ and σ from (7.3). As a result, the likelihood function depends only on θ_i and p_i for

$i = 1, \ldots, d$. (7.3) can then be maximized to obtain estimates of $\hat{\theta}_i$ and $\hat{p}_i$. The estimates $\hat{\mu}$ and $\hat{\sigma}^2$ can then readily be obtained from (7.4) and (7.5).

7.5.1.2 The Best Linear Unbiased Prediction and Predictive Distribution

Given the hyper-parameter estimates $\hat{\theta}_i$, $\hat{p}_i$, $\hat{\mu}$ and $\hat{\sigma}^2$, $y = f(x)$ at any untested point x can be predicted by using the function values y^i at x^i for $i = 1, \ldots, K$. The best linear unbiased predictor of $f(x)$ is [33, 47]:

$$\hat{f}(x) = \hat{\mu} + r^T C^{-1} (y - \mathbf{1}\hat{\mu}) \qquad (7.6)$$

and its mean squared error is:

$$s^2(x) = \hat{\sigma}^2 [1 - r^T C^{-1} r + \frac{(1 - \mathbf{1}^T C^{-1} r)^2}{\mathbf{1}^T C^{-1} r}] \qquad (7.7)$$

where $r = (c(x, x^1), \ldots, c(x, x^K))^T$. $N(\hat{f}(x), s^2(x))$ can be regarded as a predictive distribution for $f(x)$ given the function values y^i at x^i for $i = 1, \ldots, K$.

7.5.2 Discussions of GP Modeling

Some interesting issues are discussed in the following for better understanding of the GP machine learning method:

- The advantages and disadvantages of GP machine learning
  Two advantages of Gaussian models are their tractability and preciseness. In regression methods, we can consider the deterministic response $y(x)$ as a sample of the stochastic process $Y(x) = \sum_{i=1}^{k} \omega_i f_i(x) + \varepsilon(x)$ ($\varepsilon(x)$ is the regression error). Some surrogate modeling methods, such as artificial neural networks (ANNs), focus on the techniques to construct the weighted sum of a set of functions. By its transfer function, an ANN can theoretically approximate any $y(x)$ [48]. However, the method to determine the type and the number of neurons ($f_i(x)$) is not mathematically sound and much experience has to be used. Hence, the training parameters are very important and over-fitting may appear for ANNs. Ordinary Gaussian Process (GP) surrogate modeling,[1] on the other hand, simplifies the weighted sum of $f_i(x)$ to a constant μ independent of x, and the regression error $\varepsilon(x) \sim N(0, \sigma^2)$. The GP model carries out regression by analyzing the correlations between $\varepsilon(x)$ and x. Therefore, GP surrogate modeling is a theoretically principled method for determining a much smaller number of free model parameters. Besides that, an advantage of GP compared to some other popular surrogate

---

[1] There are also simple GP modeling and blind GP modeling [49].

modeling techniques is that it provides estimates of the model uncertainty [50, 51].

The drawback of the Gaussian Process surrogate modeling on the other hand is the long training time compared to other popular methods, such as ANN, RBF. The computational complexity will be elaborated in the following paragraphs and some solution methods will be discussed.

- The correlation function in GP machine learning

The correlation of two samples should be between 0 and 1. The closer the distance between two samples, the higher the correlation and viseversa. Therefore, the correlation is defined as:

$$c(x^i, x^j) = exp(-dist(x^i, x^j))$$ (7.8)

where $dist(x^i, x^j)$ is the distance between x^i and x^j. Suppose the distance is 0, which means that x^i and x^j are in fact the same point, then the correlation is 1. In contrast, if the distance between x^i and x^j is large, the correlation should approach 0.

The distance between x^i and x^j is defined as:

$$dist(x^i, x^j) = \sum_{l=1}^{d} \theta_l |x_l^i - x_l^j|^{p_l}$$ (7.9)

The coefficients θ and p allow generalizations compared to traditional distance definitions such as the Euclidean distance. The reason of using θ is that the activity of the l variables in x is different. If a small value of $|x_k^i - x_k^j|^{p_k}$ translates to large differences in the function value, the kth variable is more important (more active), then θ_k is large and the resulted correlation is low. In contrast, traditional distance definitions, which weigh the variables equally, cannot reflect the activity of different variables. The reason to use p is to reflect the smoothness of the function to be regressed. The range of p is $[1, 2]$. The smoothness increases with p. Physical phenomena likely showing a linear behavior near the origin are considered as less smooth; while those likely showing a parabolic behavior near the origin are considered as smoother (the phenomena must be continuously differentiable). There are some other correlation functions, and all of them obey the above ideas [52].

- The goal to optimize the likelihood function

Maximum-likelihood estimation is a method of estimating the parameters (in Gaussian Process, these are μ and σ^2) of a statistical model. For a fixed set of data and underlying statistical model, the method of maximum likelihood selects those values of the model parameters that produce a distribution that gives the observed data the greatest probability. Using the likelihood function of the Gaussian stochastic process (Eq. (7.3)), and setting $\frac{\partial}{\partial \mu} \log(h)$ and $\frac{\partial}{\partial \sigma^2} \log(h)$ to 0, Eq. (7.4) (μ estimate) and (7.5) (σ^2 estimate) can be derived to maximize the likelihood function h. This is solved in closed form, assuming that the vector θ and p are known. However, θ and p are parameters we want to estimate and they affect the correlation matrix R in the likelihood function. Therefore, we substitute Eqs. (7.4) and (7.5) to h (Eq. (7.3)) to construct a function depending only upon θ and p. This is the function we want to optimize in practice and the goal is to obtain the best values of θ and p.

- The methods to optimize the likelihood function
  There are many possible methods. Quasi Newton methods can be employed. Details of the application to the likelihood function in Eq. (7.3) are in [53]. However, since the problem is multimodal, it cannot be guaranteed that gradient-based methods can succeed. Direct search methods like pattern search [54], or the Nelder–Mead method (see Chap. 3) can be employed. The blind DACE toolbox [55] includes most of these methods. Blind DACE is an enhanced version of the DACE toolbox [52]. Evolutionary computation methods are also very promising. The investigation of the likelihood function optimization is important in Gaussian Process machine learning, especially when the dimensionality is high. The optimized results of θ and p directly influence the prediction quality.
- The ground of Eqs. (7.6) and (7.7)
  Equations (7.6) and (7.7) are based on the best linear unbiased prediction. Given the input x and output y, consider the linear predictor at an untried x^*.

$$\widehat{f}(x^*) = z^T(x^*)y \tag{7.10}$$

In Eq. (7.10), z is a $n \times 1$ vector. We can replace y by the corresponding random quantity, which is in the Gaussian Process model, Y. Then, $\widehat{f}(x^*)$ is a random number, whose mean squared error (MSE) of this predictor averaged over the stochastic process is:

$$MSE[\widehat{f}(x^*)] = E[z^T(x^*)Y - Y(x^*)]^2 \tag{7.11}$$

We want to minimize the MSE under the constraint of unbiasedness, which is as:

$$E[z^T(x^*)Y] = E[Y(x^*)] \tag{7.12}$$

Introducing Lagrangian multipliers, this optimization problem can be solved in closed form. Because in Gaussian Process machine learning, we only use a constant μ as the regression function, Eq. (7.6) can be derived. Eq. (7.7) is in the form of [47]:

$$MSE[\widehat{f}(x^*)] = \widehat{\sigma}^2[1 - r^T C^{-1} r] \tag{7.13}$$

The term $\frac{(1 - 1^T C^{-1} r)^2}{1^T C^{-1} r}$ in Eq. (7.7) is added to take the full estimation error into account because the used μ is estimated by the samples instead of the exact value [53].

- The computational complexity of GP machine learning
  The computational complexity of GP machine learning is $O(Nm^3 d)$, where N denotes the number of iterations spent for adjusting the parameters θ and p, d is

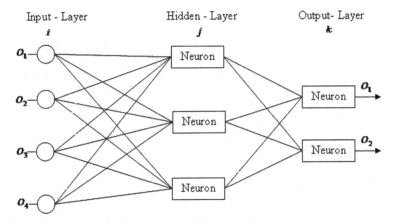

Fig. 7.3 Artificial neural network structure

the number of variables (dimensionality) of x, and m is the number of training points. It can be seen that m affects the training cost the most. However, d is also very important and N is affected by d. When d is large, a larger N is necessary to obtain a satisfactory result.

7.6 Artificial Neural Networks

An artificial neural network (ANN) is a computational mechanism, the structure of which essentially mimics the process of knowledge acquisition, information processing and organizational skills of a human brain. An ANN has the capability of learning complex nonlinear relationships and associations from a large volume of data [56]. An ANN can be used for function approximation, pattern recognition, clustering, compression, control, etc. Here, we only discuss a function approximation or regression. The regression of ANN can be considered as nonlinear weighted sum. A typical ANN structure is shown in Fig. 7.3.

An ANN is composed of a number of highly interconnected neurons, usually arranged in several layers (see Fig. 7.3). These layers generally include an input layer, a number of hidden layers and an output layer. Signals generated from the input layer propagate through the network on a layer-by-layer basis in the forward direction. Neurons in the hidden layers are used to find associations between the input data and to extract patterns that can provide meaningful outputs. The output of each neuron (see Fig. 7.4) that responds to a particular combination of inputs has an impact on the overall output. The weight is controlled by the level of the activation of each neuron, and the strength of the connections between the individual neurons. Patterns of activation and interconnections are adjusted through a training process to achieve the desired output for the training data. If the average error is within a predefined

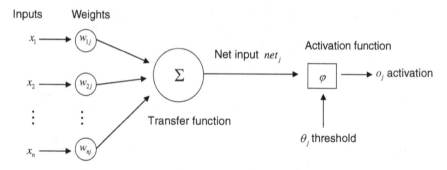

Fig. 7.4 An artificial neuron

tolerance, the training is stopped and the weights are locked in; the network is then ready to be used [48]. Theoretically, ANN can approximate any function.

7.7 Summary

In this chapter, we first have reviewed the current research of RF and mm-wave integrated components, circuit and antenna synthesis. We can draw the following conclusions:

(1) Only integrated passive components working at low frequencies can be efficiently synthesized with a good result by the traditional methods. For high-frequency (above 40 GHz) passive components, if SAEAs are not introduced, either a low-quality result may be obtained or a very long computational time may be cost.

(2) For RF and mm-wave integrated circuit synthesis, only the low-frequency RF circuits (<10 GHz) can be synthesized well using the traditional methods. It is very challenging to synthesize mm-wave integrated circuits, despite that there are local optimization methods if the designer can provide a good initial design.

(3) In antenna synthesis, although EAs are widely applied, the efficiency must be further enhanced. The a few efficient optimization methods reported can only solve small-scale problems (the designer must first extract the most critical parameters or the antenna can only be simple). For other methods, a good starting point is required from the designer.

Therefore, it can be seen that fundamental and thorough research is needed for EM simulation-based design automation in order to meet the industrial requirements. The reason for these challenges is the high computational cost of the EM simulations. In circuit synthesis, existing lumped equivalent circuit models are only applicable to low-frequency cases. Hence, a new kind of methods is necessary for high-frequency passive components, circuits and complex antennas. Then, the state-of-the-art research for the computational intelligence background of this new kind of methods, i.e., surrogate model assisted evolutionary computation (SAEA), has been discussed. A critical challenge for current SAEA research is the "curse of dimensionality", which must be addressed for mm-wave integrated circuit and complex antenna synthesis. At last, the main learning machine used in the following chapters, Gaussian Process, has been discussed. A brief introduction of ANN was also provided.

References

1. Ohmori S, Yamao Y, Nakajima N (2000) The future generations of mobile communications based on broadband access technologies. IEEE Commun Mag 38(12):134–142
2. Niknejad A, Hashemi H (2008) mm-Wave silicon technology: 60 GHz and beyond. Springer, New York
3. Nieuwoudt A, Massoud Y (2006) Variability-aware multilevel integrated spiral inductor synthesis. IEEE Trans Comput-Aided Des Integr Circuits Syst 25(12):2613–2625
4. Mandal S, Goyel A, Gupta A (2009) Swarm optimization based on-chip inductor optimization. In: Proceedings of 4th international conference on computers and devices for, communication, Kolkata, India, pp 1–4 December 2009
5. Mandal S, Sural S, Patra A (2008) ANN-and PSO-based synthesis of on-chip spiral inductors for RF ICs. IEEE Trans Comput-Aided Des Integr Circuits Syst 27(1):188–192
6. Koziel S (2009) Surrogate-based optimization of microwave structures using space mapping and kriging. In: Proceedings of European microwave conference, pp 1062–1065
7. Bandler J, Cheng Q, Dakroury S, Mohamed A, Bakr M, Madsen K, Sondergaard J (2004) Space mapping: the state of the art. IEEE Trans Microw Theory Tech 52(1):337–361
8. Koziel S, Bandler J (2008) Space mapping with multiple coarse models for optimization of microwave components. IEEE Microw Wire Compon Lett 18(1):1–3
9. Rayas-Sánchez J (2004) EM-based optimization of microwave circuits using artificial neural networks: the state-of-the-art. IEEE Trans Microw Theory Tech 52(1):420–435
10. Koziel S, Bandler J (2007) Space-mapping optimization with adaptive surrogate model. IEEE Trans Microw Theory Tech 55(3):541–547
11. Allstot D, Choi K, Park J (2003) Parasitic-aware optimization of CMOS RF circuits. Springer, Netherlands
12. Choi K, Allstot D (2006) Parasitic-aware design and optimization of a CMOS RF power amplifier. IEEE Trans Circuits Syst I Regul Pap 53(1):16–25
13. Tulunay G, Balkir S (2008) A synthesis tool for CMOS RF low-noise amplifiers. IEEE Trans Comput-Aided Des Integr Circuits Syst 27(5):977–982
14. Ramos J, Francken K, Gielen G, Steyaert M (2005) An efficient, fully parasitic-aware power amplifier design optimization tool. IEEE Trans Circuits Syst I Regul Pap 52(8):1526–1534
15. Agarwal A, Vemuri R (2005) Layout-aware RF circuit synthesis driven by worst case parasitic corners. In: Proceedings of IEEE international conference on computer design: VLSI in computers and processors, pp 444–449
16. Zhang G, Dengi A, Rohrer R, Rutenbar R, Carley L (2004) A synthesis flow toward fast parasitic closure for radio-frequency integrated circuits. In: Proceedings of the 41st annual design automation conference, pp 155–158
17. De Ranter C, Van der Plas G, Steyaert M, Gielen G, Sansen W (2002) CYCLONE: automated design and layout of RF LC-oscillators. IEEE Trans Comput-Aided Des Integr Circuits Syst 21(10):1161–1170
18. Chu M, Allstot D (2005) Elitist nondominated sorting genetic algorithm based RF IC optimizer. IEEE Trans Circuits Syst I Regul Pap 52(3):535–545
19. Park J, Choi K, Allstot D (2003) Parasitic-aware design and optimization of a fully integrated CMOS wideband amplifier. In: Proceedings of Asia and South Pacific design automation conference, pp 904–907

20. Hoorfar A (2007) Evolutionary programming in electromagnetic optimization: a review. IEEE Trans Antennas Propag 55(3):523–537
21. Rocca P, Oliveri G, Massa A (2011) Differential evolution as applied to electromagnetics. IEEE Antennas Propag Mag 53(1):38–49
22. Robinson J, Rahmat-Samii Y (2004) Particle swarm optimization in electromagnetics. IEEE Trans Antennas Propag 52(2):397–407
23. Pantoja M, Bretones A, Martín R (2007) Benchmark antenna problems for evolutionary optimization algorithms. IEEE Trans Antennas Propag 55(4):1111–1121
24. Liu W (2005) Design of a multiband CPW-fed monopole antenna using a particle swarm optimization approach. IEEE Trans Antennas Propag 53(10):3273–3279
25. Yeung S, Man K, Luk K, Chan C (2008) A trapeizform U-slot folded patch feed antenna design optimized with jumping genes evolutionary algorithm. IEEE Trans Antennas Propag 56(2):571–577
26. Goudos S, Siakavara K, Samaras T, Vafiadis E, Sahalos J (2011) Self-adaptive differential evolution applied to real-valued antenna and microwave design problems. IEEE Trans Antennas Propag 99:1286–1298
27. Gregory M, Bayraktar Z, Werner D (2011) Fast optimization of electromagnetic design problems using the covariance matrix adaptation evolutionary strategy. IEEE Trans Antennas Propag 59(4):1275–1285
28. Jin N, Rahmat-Samii Y (2007) Advances in particle swarm optimization for antenna designs: Real-number, binary, single-objective and multiobjective implementations. IEEE Trans Antennas Propag 55(3):556–567
29. Goudos S, Zaharis Z, Kampitaki D, Rekanos I, Hilas C (2009) Pareto optimal design of dual-band base station antenna arrays using multi-objective particle swarm optimization with fitness sharing. IEEE Trans Magn 45(3):1522–1525
30. John M, Ammann M (2009) Antenna optimization with a computationally efficient multiobjective evolutionary algorithm. IEEE Trans Antennas Propag 57(1):260–263
31. Knowles J (2006) ParEGO: a hybrid algorithm with on-line landscape approximation for expensive multiobjective optimization problems. IEEE Trans Evol Comput 10(1):50–66
32. Siah E, Sasena M, Volakis J, Papalambros P, Wiese R (2004) Fast parameter optimization of large-scale electromagnetic objects using DIRECT with kriging metamodeling. IEEE Trans Microw Theory Tech 52(1):276–285
33. Jones D, Schonlau M, Welch W (1998) Efficient global optimization of expensive black-box functions. J Global optim 13(4):455–492
34. Jin Y (2005) A comprehensive survey of fitness approximation in evolutionary computation. Soft Comput-A Fus Found Methodol Appl 9(1):3–12
35. Emmerich M, Giannakoglou K, Naujoks B (2006) Single-and multiobjective evolutionary optimization assisted by gaussian random field metamodels. IEEE Trans Evol Comput 10(4):421–439
36. Giannakoglou K (2002) Design of optimal aerodynamic shapes using stochastic optimization methods and computational intelligence. Prog Aerosp Sci 38(1):43–76
37. Ong Y, Lum K, Nair P (2008) Hybrid evolutionary algorithm with hermite radial basis function interpolants for computationally expensive adjoint solvers. Comput Optim Appl 39(1):97–119
38. Zhang Q, Liu W, Tsang E, Virginas B (2010) Expensive multiobjective optimization by MOEA/D with gaussian process model. IEEE Trans Evol Comput 14(3):456–474
39. Voutchkov I, Keane A (2010) Multi-objective optimization using surrogates. Comput Intell Optim 7:155–175
40. Jin Y, Olhofer M, Sendhoff B (2002) A framework for evolutionary optimization with approximate fitness functions. IEEE Trans Evol Comput 6(5):481–494
41. Jones D (2001) A taxonomy of global optimization methods based on response surfaces. J Global Optim 21(4):345–383
42. Liu B, Zhao D, Reynaert P, Gielen G (2011) Synthesis of integrated passive components for high-frequency RF ICs based on evolutionary computation and machine learning techniques. IEEE Trans Comput-Aided Des Integr Circ Syst 30(10):1458–1468

43. Lim D, Jin Y, Ong Y, Sendhoff B (2010) Generalizing surrogate-assisted evolutionary computation. IEEE Trans Evol Comput 14(3):329–355
44. Zhou Z, Ong Y, Nair P, Keane A, Lum K (2007) Combining global and local surrogate models to accelerate evolutionary optimization. IEEE Trans Syst Man Cybern Part C Appl Rev 37(1):66–76
45. Liu B, Zhang Q, Gielen G (2013) A gaussian process surrogate model assisted evolutionary algorithm for medium scale expensive black box optimization problems. IEEE, Trans Evol Comput (To be published)
46. Rasmussen C (2004) Gaussian processes in machine learning. Adv Lect Mach Learn 3176:63–71
47. Sacks J, Welch W, Mitchell T, Wynn H (1989) Design and analysis of computer experiments. Stat Sci 4(4):409–423
48. Wasserman P (1989) Neural computing: theory and practice. Van Nostrand Reinhold Co. 21(3):2–7, Sep 1989
49. Couckuyt I, Forrester A, Gorissen D, De Turck F, Dhaene T (2012) Blind kriging: implementation and performance analysis. Adv Eng Softw 49:1–13
50. Ackermann E, De Villiers J, Cilliers P (2011) Nonlinear dynamic systems modeling using gaussian processes: predicting ionospheric total electron content over south africa. J Geophys Res 116:A10,303
51. Buche D, Schraudolph N, Koumoutsakos P (2005) Accelerating evolutionary algorithms with gaussian process fitness function models. IEEE Trans Syst Man Cybern Part C Appl Rev 35(2):183–194
52. Lophaven S, Nielsen H, Søndergaard J (2002) DACE: A Matlab kriging toolbox. Citeseer. http://www2.imm.dtu.dk/~hbni/dace
53. Krishnaiah P (1988) Handbook of statistics. Motilal Banarsidass Publishers, Springer Netherlands, North-Holland
54. Dennis J, Torczon V (1997) Managing approximation models in optimization. In: Multidisciplinary design optimization: State-of-the-art. SIAM, Philadelphia, pp 330–347
55. Gorissen D, Couckuyt I, Demeester P, Dhaene T, Crombecq K (2010) A surrogate modeling and adaptive sampling toolbox for computer based design. J Mach Learn Res 11:2051–2055
56. Javadi A (2000) Estimation of air losses from tunnels driven under compressed air using neural networks. In: Proceedings of the 10th international conference on computer methods and advances in geomechanics, pp 207–212

Chapter 8
Passive Components Synthesis at High Frequencies: Handling Prediction Uncertainty

In the previous chapter, the background for the third part of the book, simulation-based electromagnetic (EM) design automation, has been reviewed. From this chapter on, the problems listed in Chap. 7 will be addressed one by one. The common goal of the proposed methods is to approximate the globally optimal solution or Pareto front efficiently.

This chapter will present efficient solution methods for single- and multi-objective integrated passive component (e.g., inductor, transformer) synthesis at high frequencies. Generally speaking, these problems are small-scale computationally expensive optimization problems, which is the foundation of introducing surrogate model assisted evolutionary algorithms (SAEAs). SAEAs adaptively integrate machine leaning into evolutionary computation, so as to enhance the efficiency drastically. However, when using prediction, prediction uncertainty is unavoidable, and the method to balance the correct convergence and efficiency becomes essential in SAEA. This is also called the "curse of uncertainty" in the computational intelligence field. This chapter will concentrate on the methods to handle the prediction uncertainty. Three practical SAEAs are used as examples to show the methods to construct an SAEA and to handle the uncertainty in it. Two efficient single-objective global optimization methods (GPDECO, MMLDE) and an efficient multi-objective optimization method (GPMOOG) will be presented. GPDECO and GPMOOG address the "curse of uncertainty" in a forward way by intelligently selecting prediction or exact function evaluation. They are prediction-based methods as described in Chap. 7. MMLDE, on the other hand, addresses the "curse of uncertainty" in a backward way. It utilizes the uncertainty for space exploration and changes the "curse of uncertainty" to a "bless of uncertainty" by introducing prescreening. It also integrates prediction with prescreening. Hence, MMLDE is a hybrid prescreening and prediction-based method.

The remainder of the chapter is organized as follows. Section 8.1 discusses the individual threshold control method with the practical algorithm, GPDECO, for high-frequency integrated passive component synthesis. Section 8.2 scales up GPDECO, and provides experimental results. Section 8.3 introduces the prescreening methods.

B. Liu et al., *Automated Design of Analog and High-frequency Circuits*,
Studies in Computational Intelligence 501, DOI: 10.1007/978-3-642-39162-0_8,
© Springer-Verlag Berlin Heidelberg 2014

The hybridization of prescreening and prediction is discussed in Sect. 8.4 with a practical algorithm, MMLDE. Section 8.5 introduces the generation control method to handle the prediction uncertainty. The method to handle multiple objectives in the MOEA/D framework is introduced in Sect. 8.6, together with an example, the GPMOOG algorithm and its experimental results. Section 8.7 summarizes this chapter.

In this chapter, the CI techniques are introduced together with the application background. We will use equivalently the terms "expensive exact function evaluation" and "EM simulation" depending if we look at the problem from the CI or the EDA point of view, respectively.

8.1 Individual Threshold Control Method

8.1.1 Motivations and Algorithm Structure

Chapter 7 discussed four kinds of methods for integrated passive component synthesis. They are:

(1) equivalent circuit model and global optimization algorithm based (ECGO) methods [1, 2];
(2) EM-simulation and global optimization algorithm based (EMGO) methods [3];
(3) off-line surrogate model, EM-simulation and global optimization algorithm based (SEMGO) methods [4];
(4) surrogate model and local optimization algorithm based (SMLO) methods [5, 6].

Three conclusions can be drawn from the available methods:

(1) Global optimization and EM simulations are essential to obtain high-quality solutions.
(2) Using a surrogate model to predict the performances is critical to enhance the efficiency.
(3) The reason why ECGO and SMLO need an accurate coarse model is that finding the optimal solution depends critically on the accuracy of the coarse model.

An SAEA for the targeted problem works as follows. It performs global optimization of the mm-wave structure using EM simulations, and a Gaussian process (GP)-based surrogate model is constructed and updated on-line to predict the results of future expensive EM simulations. Note that this surrogate model is not like the surrogate model used in SMLO. This surrogate model is constructed and updated during the optimization (i.e., on-line) by using the data from EM simulations, whereas the surrogate model in SMLO is constructed off-line using the data from coarse model evaluations.

Then, we can ask ourselves the following question: How to integrate the on-line surrogate model into global optimization algorithms? The answer, however, is not trivial. The challenge is that the quality of the surrogate model is improving gradually, as more training data are provided in the optimization process. Hence, to maintain the solution quality, some predicted performances cannot be used. But how to recognize if a prediction can be used remains an open question. Towards this goal, there are several possible methods. This section will introduce the individual threshold control method. This method first determines a threshold of uncertainty to decide whether the prediction from the GP model can be used for a candidate or whether the expensive fine model evaluation is necessary. For many applications, there is often a computationally cheap, but not accurate, coarse model. The threshold value can be obtained by analyzing the data of the coarse model. An algorithm, GPDECO [7], will be presented in the following section based on this idea. Unlike SMLO, GPDECO does not use the optimal solution of the coarse model to define the search space and also does not use the coarse model performance-based surrogate model to guide the search, so the requirement on accuracy of the coarse model is much lower.

8.1.2 Determination of the MSE Thresholds

The accuracy and reliability of the GP model is gradually improved in the optimization process, as additional training data are generated in the optimization process. This leads to a problem that some data predicted by the GP model may have large differences compared with the fine EM simulation, especially at the beginning of the optimization when few training data are available, and cannot be used. Fortunately, by using GP, the uncertainty measurement s^2 or MSE (Eq. 7.7) of each prediction can be used as a judging instrument. If the s^2 value for a prediction is higher than a certain threshold, we discard the predicted value and fine simulation will be performed; otherwise, it means that the error is acceptable, so the prediction can be used in the evolution process. This is shown in Fig. 8.1. But how to decide on the value of the threshold is a problem. A large threshold will cause inaccurate predictions to become the fitness value of some individuals, which may mislead the direction of

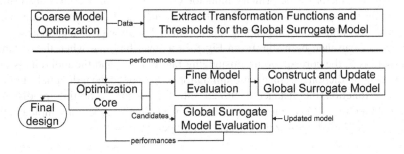

Fig. 8.1 General idea of GPDECO

evolution or even make it fail. A small threshold can maintain the accuracy, but it will also discard predictions that could be used, so the efficiency is not enhanced much. A feasible method is to extract the s^2 thresholds by observing the data generated by the coarse model. Although the coarse model is not accurate, it represents the same problem as the fine EM model. Hence, we can reasonably assume that the input/output ranges, the severity of the constraints and the convergence process are similar. Therefore, using the data from the coarse model to extract the thresholds is reliable.

To achieve this, first the coarse model is used to perform the optimization with the same objective function and constraints. After each iteration, the GP model is trained based on the available data and the performances of the candidate solutions in the next iteration are predicted. Note that the predicted performances are not used in the optimization in this step. After this step, we have all the individuals that appeared along the coarse model optimization and two performance values for each of them: one provided by the coarse model evaluator and one predicted by the GP model, which are the source to extract the thresholds. Note that the fine EM model is not considered in this step, and the surrogate model trained by coarse model performances is not used for the real synthesis of the microwave structure. In addition, this step is often cheap, because the coarse models are often quite simple. For example, an equivalent circuit can be used as the coarse model for a microwave device.

Then, the selection of the s^2 thresholds can be split into two trade-off problems. First, the distance thresholds between the values predicted by GP and the values by the coarse model simulation should be considered. If the distance is within the threshold, the predicted values will be used. The larger the distance threshold values, the higher the inaccuracy of the prediction. The smaller the distance thresholds values, the less predictions that can be used. Hence, there is a trade-off. For constraints, we use the misjudging rate to reflect the accuracy. The number of misjudgments is the sum of individuals whose predicted values satisfy a constraint but for which the true simulation value does not, or viceversa. The misjudging rate is the ratio of the number of misjudgments over the total number of individuals. Figure 8.2a shows two percentages with different distance thresholds, which are the misjudging rate and ratio of the number of prediction used over the sum of predictions and exact function evaluations (we call the latter prediction rate). The used example is a constraint of the quality factor of the primary inductor $Q_1 > 10$ for an overlay transformer. According to experiments with different transformers and different constraints, a distance threshold with a misjudging rate less than 2 % often gives a good accuracy. A trade-off can then be made. From Fig. 8.2a we can find that when the distance threshold is 5, the corresponding misjudging rate is 1.2 % and the individuals that can use GP prediction to replace fine simulation is 94 % of all the individuals. For the objective function, the distance itself can reflect the accuracy. Setting the distance threshold within 1–3 % of the average value of the objective function sampled in the coarse model-based optimization process often performs well.

After the distance thresholds are decided, we can connect the s^2 value with the distance values described above. The MSE (or s^2) value provided by GP prediction for each candidate is an estimation of the uncertainty. This can generally reflect the distance between the predicted value and the exact value described above. Under a determined distance threshold t_d, we can sweep the s^2 threshold, t_s. For a certain value of t_s in the sweeping, there are some individuals whose s^2 values are within the t_s, and the distances between the predicted values and the exact values for them are also within the given t_d, but there also exist some individuals that are opposite, i.e., the distances between the predicted values and the exact values are larger than t_d. Both percentages (the kinds of points over the total number of individuals) can be calculated for a t_s value in the sweeping. We want to maximize the first percentage and minimize the second percentage. A trade-off can be made by exploring the two corresponding percentages for different values of t_s. Figure 8.2b shows the quality factor of the primary inductor Q_1 for an overlay transformer under a determined distance threshold. According to experiments, the second percentage is suggested to be within 10 % to maintain the reliability of the mapping and a trade-off can then be made to decide t_s.

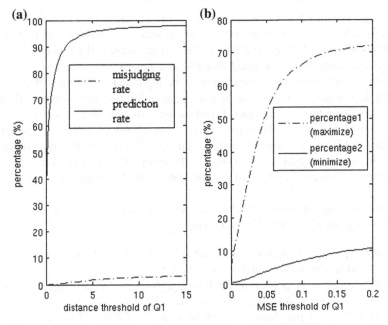

Fig. 8.2 Two trade-offs to decide the s thresholds (Q_1, example in Sect. 8.2)

8.2 The GPDECO Algorithm

8.2.1 Scaling Up of GPDECO

An SAEA, called Gaussian Process-Based Differential Evolution for Constrained Optimization (GPDECO) is proposed based on the individual threshold control for high-frequency integrated passive component synthesis, which aims to: (1) achieve comparable results with EMGO (the best algorithm with respect to the solution quality aspect) in high-frequency synthesis, (2) dramatically improve the efficiency of EMGO and make the computational time practical.

The DE algorithm [8] is selected as the optimization engine. The synthesis of microwave structures is a constrained optimization problem. Therefore, different constraint handling methods introduced in Chaps. 2 and 3 can be applied. In GPDECO, the selection-based method for constraint handling introduced in Chap. 2 is used.

With the on-line constructed GP model and the decided s^2 thresholds, the surrogate model, whose predictions can be judged to decide if the model predictions can be used or EM simulations are needed, can be integrated into the optimization flow. However, two additional considerations are necessary:

(1) Even when the GP model predicts the performances well (within the MSE thresholds), there still exists a small error, so the final result must be verified by the EM simulation. Hence, in GPDECO, the best candidate in the population in each generation needs to be verified with EM simulations.
(2) GP assumes a Gaussian distribution of the input and output data. Hence, normalization should be done on the training data. But for some problems, directly normalizing the design points X and the performances Y may not yield good results. For example, the upper bound of some design variable may be 10–20 times larger than its lower bound. A solution is to use transformation functions (e.g. $log(x)$) to decrease the range of the input (X)/output (Y) variables. The output values can be analyzed with the data of the coarse model.

The flow diagram of the GPDECO algorithm is shown in Fig. 8.3.
The algorithm consists of the following steps.

Step 0:   Optimize the coarse model and extract transformation functions and the MSE thresholds as described in Sect. 8.1.2.

Step 1:   Initialize the population randomly, perform fine model evaluations of the individuals and construct the GP model.

Step 2:   For a new population of candidates, use the GP model to predict the performances and the corresponding s^2 values.

Step 3.1: For the individuals whose s^2 values satisfy the thresholds, use the predicted values and go to step 4.

Step 3.2: For the individuals whose s^2 values do not satisfy the thresholds, use fine EM model evaluations. Add these data to the training data set and update the GP model. Go to step 4.

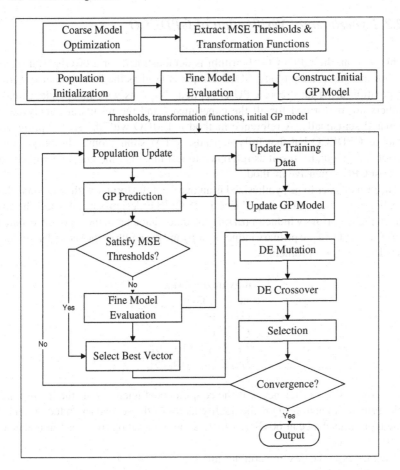

Fig. 8.3 Flow diagram of GPDECO

Step 4: Select the current best candidate using the tournament selection-based method. If it has not been evaluated by fine EM model, use fine EM model to evaluate it and update the GP model.

Step 5: Perform the DE mutation and crossover operations to obtain each individual's trial individual. Predict or evaluate the trial individuals using the techniques in Step 3.

Step 6: Perform selection between each individual and its corresponding trial counterpart according to selection rules in the tournament selection-based method (see Chap. 2) to update the population.

Step 7: If the stopping criterion is met (e.g., a convergence criterion or a maximum number of iterations), then output the best candidate found so far and its performance values; otherwise go back to Step 2.

8.2.2 Experimental Verification of GPDECO

In this section, the GPDECO algorithm is demonstrated for a 60 GHz transformer in a 90 nm CMOS technology. SONNET [9] is used as the EM simulator. Parallel computation using an 8-core CPU is applied. Ten runs with independent random numbers are performed for all the experiments and the results are analyzed and compared statistically. A reference method is EMGO with the same optimization kernel as GPDECO but that fully depends on EM simulations. The purpose is to provide a high quality result as reference to test the other methods. Obviously, it is the most CPU expensive method.

The transformer to be synthesized is an overlay transformer with octagonal shape (See Fig. 8.4). The output impedance is 25 Ω. The design variables are the inner diameter of the primary inductor ($dinp$), the inner diameter of the secondary inductor ($dins$), the width of the primary inductor (wp) and the width of the secondary inductor (ws). The synthesis problem is:

$$
\begin{aligned}
&\text{maximize } Gmax \\
&\text{s.t. } k > 0.85 \\
&Q_1 > 10 \\
&Q_2 > 10 \\
&Re(Z_{in}) \in [10, 20] \\
&Im(Z_{in}) \in [10, 25]
\end{aligned}
\tag{8.1}
$$

where $Gmax$ is the efficiency, k is the coupling coefficient, Q_1 is the quality factor of the primary inductor, Q_2 is the quality factor of the secondary inductor, and Z_{in} is the input impedance at 60 GHz, which is the required optimal load impedance of the driver stage.

The ranges of the design variables are:

$$
\begin{aligned}
&dinp, dins \in [20, 150] \, \mu m \\
&wp, ws \in [5, 10] \, \mu m
\end{aligned}
\tag{8.2}
$$

The results are shown in Table 8.1, where $Gmax$ is the average value of the 10 runs, RCS is the number of designs satisfying the constraints over 10 runs and N is the number of evaluations.

Table 8.1 Results of GPDECO and EMGO

Item		EMGO	GPDECO
RCS		10/10	10/10
Gmax		86.1 %	86.1 %
Average	N	979	202
Average	time	1.4 h	0.3 h

Fig. 8.4 A typical result

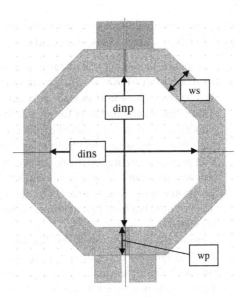

From the optimized *Gmax* and the number of evaluations in Table 8.1, it can be concluded that the result of GPDECO has no loss compared with the benchmark (EMGO) method, but it is nearly five times faster. GPDECO costs 0.3 h. A typical result of GPDECO is shown in Fig. 8.4. The design solutions of both GPDECO and EMGO are close to $[dinp, dins, wp, ws] = [54.7, 54.8, 10, 10]$ μm.

From the experiment, it can be seen that GPDECO solves the high dependence on the (limited) accuracy of the coarse model. The example shows that GPDECO can synthesize high-performance transformers. GPDECO has similar quality of results compared to the EMGO framework, which is a promising framework in terms of solution quality. Yet GPDECO only costs 20–25 % of the computational effort, and therefore makes the synthesis have a high efficiency.

However, the main limitation of GPDECO is that an equivalent circuit model of the passive component is still necessary for the determination of the MSE threshold, although the results have a low sensitivity to its accuracy. In some cases, the equivalent circuit models cannot be found. In such cases, GPDECO cannot be applied.

8.3 Prescreening Methods

8.3.1 The Motivation of Prescreening

For an EA using surrogate model, the purpose of the surrogate model is to approximate the hyper-surface of the function to be optimized in order to replace expensive exact function evaluations by predictions of the surrogate model. To achieve this goal, there are several kinds of methods.

1. The easiest but suboptimal one is to use an off-line surrogate model, which is the approach in SEMGO [4]. Such method first constructs a good surrogate model which covers the whole design space and then uses it. No matter whether there is surrogate model updating for local refinement in the optimization process, the global exploration is supported by the off-line surrogate model. Hence, to obtain a reliable surrogate model, the training data need to cover the whole design space with a reasonably high density. Hence, a lot of EM simulations are necessary. On the other hand, only a small part of the design space is typically useful in the optimization. The reason is that the optimization algorithm is not based on enumeration, but based on iteration, so many of these expensive EM simulations are wasted.

2. A more efficient way is to use on-line surrogate models. SAEA holds the idea: "in the deep darkness of the design space, there is no need to lighten the whole world but rather the close vicinity of the path to the destination". An SAEA therefore first constructs a very rough surrogate model, and then improves it on line but only in the necessary area of the design space, as determined by the optimization algorithm and the updating technique. Then, a question arises: which area of the design space is necessary? Two methods have been developed in CI: prediction and prescreening.

   a. Prediction methods concentrate on the areas of the design space visited by the optimization engine. For example, the generated candidates in an EA may either be in the promising regions or non-promising regions. Both of them are considered to be useful for prediction-based methods. GPDECO provides a way to construct a surrogate model based on the prediction method. No matter a candidate solution is promising or not, the surrogate model should provide a reliable prediction to it. In this way, there is a high probability that the correct optimal solutions can be found, since it follows a traditional EA search route and only reliable predictions are used. To that end, prediction methods can be considered as a conservative but safe method. On the other hand, for the areas of the design space that are not visited by the EA, prediction methods ignore them, but a lot of computational effort may be spent in them when using off-line surrogate model.

   b. Prescreening methods concentrate on areas of the design space that may have promising solutions according to the current evaluated data. Prescreening methods utilize the prediction uncertainty, and pay attention to the candidates with good prediction value and less prediction uncertainty or those with reasonably good prediction value and large prediction uncertainty. In this way, a promising candidate may be "guessed" without a reliable surrogate model around it, while prediction-based methods often need more expensive function evaluations. However, there must exist conditions that the "guessing" is wrong, since a candidate with reasonably good prediction value and large prediction uncertainty may either be a good one or a bad one in reality. To that end, prescreening methods can be considered as a high reward but risky method. However, various empirical studies show that for small-scale single-objective optimization problems, prescreening methods show higher efficiency and comparable solution qualities compared to prediction-based methods.

8.3.2 Widely Used Prescreening Methods

The four commonly-used prescreening methods are: the most likely improvement (MI) [10], the probability of improvement (PI) [11], the expected improvement (EI) [12] and the lower confidence bound (LCB) [13]. An intensive research has been carried out in [10] based on the $(\mu + \lambda)$-ES framework on 20-dimensional problems. Note that MI does not utilize the prediction uncertainty, and it is not a typical prescreening method. Sometimes, the prescreening methods mainly refer to just EI, PI and LCB. Note that EI, PI and LCB are only suitable for GP-based surrogate modeling, since they are also based on the Gaussian stochastic process.

Considering a minimization problem, the four existing prescreening methods, MI, EI, PI, and LCB, are defined as follows:

(1) Most likely improvement (MI):

$$M(I(x)) = \begin{cases} 0, & \text{if } \widehat{y}(x) > f_{min} \\ f_{min} - \widehat{y}(x), & \text{otherwise} \end{cases} \tag{8.3}$$

where f_{min} is the current best function value in the population and $\widehat{y}(x)$ is the predicted value of a candidate. $I(x)$ represents the function of improvement compared to f_{min}.

(2) Expected improvement (EI):

$$E[I(x)] = (f_{min} - \widehat{y}(x)) \, \Phi\left(\frac{f_{min} - \widehat{y}(x)}{\widehat{s}(x)}\right) + \widehat{s}(x)\phi\left(\frac{f_{min} - \widehat{y}(x)}{\widehat{s}(x)}\right) \tag{8.4}$$

where $\phi(\cdot)$ is the standard normal density function, and $\Phi(\cdot)$ is the standard normal distribution function and $\widehat{s}(x)$ is prediction variance of the Gaussian model.

(3) Probability of improvement (PI):

$$P[I(x)] = \Phi\left(\frac{f_{min} - \widehat{y}(x)}{\widehat{s}(x)}\right) \tag{8.5}$$

(4) Lower confidence bound (LCB):

$$f_{lb}(x) = \widehat{y}(x) - \omega\widehat{s}(x), \, \omega \in [0, 3] \tag{8.6}$$

The EI, PI and LCB methods reward the prediction uncertainty in different ways. The PI method rewards candidates with a possibly small improvement but with a high probability ($\widehat{s}$ is small) and candidates with a possibly large improvement but with a low probability ($\widehat{s}$ is large). Moreover, a larger $\widehat{s}$ value is only rewarded when the predicted value of a candidate is worse than the current best function value. The EI values, on the other hand, first decrease and then grow again with increasing $\widehat{s}$ [10]. The LCB always rewards $\widehat{s}$ to a fixed extent. Although these three methods have good exploration ability, they may select individuals which are not really good but have a large $\widehat{s}$ value. The extent to reward the prediction uncertainty is in fact a problem of balancing exploration and exploitation. A larger rewarding extent emphasizes exploration, and a smaller extent emphasizes exploitation. The EI, PI and LCB methods have been compared in [10] and they show similar performances in the experiments. It is known that the LCB uses an optimistic rewarding criterion and the EI and PI methods consider the balance between exploration and exploitation. Hence, the balance between exploration and exploitation of the EI and PI methods does not work better than the optimistic rewarding criterion of LCB from the result of [10]. In the following, we will use EI as an example to show how rewarding the prediction uncertainty works.

According to (8.4), EI is the part of the curve of standard error ($\widehat{s}$) in the model that lies below the best function value sampled so far. As shown in the example of Fig. 8.5, the function value at $x = 8$ is better than that at $x = 3$, but $x = 8$ cannot be selected when the GP prediction values are used. However, point $x = 8$ can be selected when using the EI prescreening method. In Fig. 8.6, the probability density function of the prediction uncertainty at $x = 8$ is represented by curve B. We find that at the tail of the density function (area A), the EI value of $x = 8$ is better than the EI of the current f_{min} (near $x = 3$), so it is possible that the true value at $x = 8$ is better than the current f_{min}. This example shows the advantage of quickly exploring promising areas when using prescreening-based methods for small-scale single-objective expensive optimization problems. When using prediction-based methods (just the predicted values are used), the candidate $x = 8$ is likely to be selected only when there are

Fig. 8.5 The *solidline* represents an objective function that has been sampled at the five points shown as *dots*. The *dottedline* is a DACE predictor [14] fit to these points (from [12])

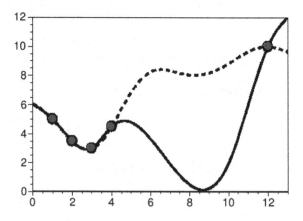

Fig. 8.6 The uncertainty about the function's value at a point (such as $x = 8$ above) can be treated as if there were a realization of a normal random variable with mean and standard deviation given by the DACE predictor [14] and its standard error (from [12])

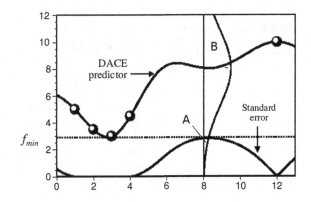

enough samples around it, but this implies more expensive exact function evaluations than using prescreening.

8.4 MMLDE: A Hybrid Prescreening and Prediction Method

MMLDE is a method that uses both prescreening and prediction for high-frequency integrated passive component synthesis [15]. The EI prescreening and the ANN-based prediction are used in MMLDE. The main advantages of MMLDE over GPDECO are:

(1) MMLDE does not need any coarse model nor the complex tuning of the parameters.
(2) MMLDE has a higher efficiency.

Compared with EMGO, MMLDE can obtain comparable results, and has approximately a tenfold improvement in computational efficiency, compared with four to five times speed enhancement of GPDECO. The speed/quality trade-off of MMLDE in the synthesis of integrated passive components at high frequencies vs. the other methods is shown in Fig. 8.7.

8.4.1 General Overview

A revised DE algorithm is used in MMLDE as the optimization kernel. Although a group of candidates is generated in each iteration, only one EM simulation is performed for the candidate with the possible best potential. Following the mutation and crossover operation, the selection operation decides on the population of the next generation $(t + 1)$. In standard DE, the trial individual after mutation and crossover, $U_i(t + 1)$ is compared to the initial target candidate solution $x_i(t)$ by means of a

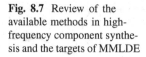

Fig. 8.7 Review of the available methods in high-frequency component synthesis and the targets of MMLDE

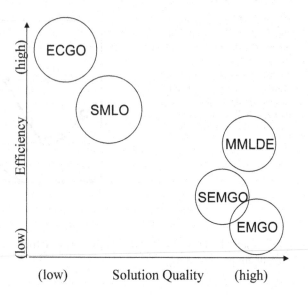

one-to-one greedy selection criterion. However, in MMLDE, that selection operator is not used, because the number of EM simulations needs to be minimized. Instead, the best solution (or the solution with the possible best potential) among all the trial solutions $U(t + 1)$ is selected and then EM simulation is applied only to this solution. Then, this new solution is added to the population. Note that in MMLDE the "population" only refers to the samples with EM simulation results, and a new point is added into the population at each iteration.

In MMLDE, after a small number of Latin-Hypercube samples (LHS) [16] of the design space are generated and EM simulations are used to evaluate these samples, the initial surrogate model is trained as a rough estimation of the performances of the microwave structure. In each iteration of the SAEA, the candidate solutions are generated by the DE operators, whose performances are prescreened by EI emphasizing exploration or predicted by ANN emphasizing exploitation, in order to select the one with the best potential. The surrogate model is then updated by including the new candidate with its EM simulation result.

8.4.2 Integrating Surrogate Models into EA

On the one hand, the number of samples in the initial surrogate model needs to be as small as possible; otherwise the efficiency will dramatically decrease. On the other hand, the design space must be covered as much as possible, because too sparse samplings make very little information is available in some areas and the reliability of the surrogate model will be poor. To make a good trade-off, LHS sampling [16] is used in MMLDE. In MMLDE, the number of samples is correlated to the dimension of the design variables, d. Empirical studies show that using around $10\,d$ initial samples is a good choice.

Although the initial surrogate model can roughly reflect the performances of the RF passive components, its quality is not good enough. Therefore, if just the predicted values are used to guide further search without any prediction-based methods (e.g., generation control, which will be discussed in Sect. 8.5), it would be easily trapped in a local optimum. The interpretation is quite intuitive. For the GP model the predicted value of a point is decided by the function values of the points near it. It may be possible that there are more points around a locally optimal point than around the globally optimal point. In this case, the function value of the locally optimal point can be predicted quite well and with less uncertainty, while the function value of the globally optimal point may be predicted poorly (as the information is little) and have a large uncertainty. The result may be that the predicted value of the globally optimal point is worse than that of the locally optimal point, which will cause a wrong selection. To address this problem, the EI prescreening is used to measure the potential, which considers both the predicted value and the uncertainty of the prediction.

EI considers both global search and local search. Especially when the sampling is sparse (large standard error), it has a high ability to consider the potential for global search. It can also exploit the potential for local search, especially when the sampling is dense. However, in the MMLDE mechanism the sampling can seldom be dense, because a limited number of EM simulations must be used due to practical time constraints. Hence, the local search ability of the EI prescreening is comparably weak. In the revised DE algorithm used in MMLDE, the same mutation and crossover operators as those in Chap. 2 are used. Hence, the global or local search property is defined by the method to determine the potential of a candidate and the corresponding surrogate model type. It is like a memetic algorithm, but is controlled by the surrogate models, rather than the evolutionary operators. For the GP model, EI is used to measure the potential when focusing on global exploration; for the ANN model, the predicted value is used to measure the potential when focusing on local exploitation. An illustrative example is shown in Fig. 8.8. The function under consideration is $y = 2.5 \times sin(x)$. From the ANN prediction values and the potential measured by EI, it can be seen that EI predicts that $x(B)$ (the corresponding x value of point (B) has the best potential, but the ANN prediction correctly selects the best point $x(A)$. It can be noticed that there are two training data on each side of $x(A)$ and $x(B)$, which influence the GP prediction and the EI prediction mostly. $x(B)$ has a smaller distance to the training point with f_{min} compared with $x(A)$, and the opposite on the other side. Hence, the EI value of $x(B)$ is larger. On the other hand, in this local area without dense sampling, the ANN catches the general trend of the function by its modeling mechanism and predicts better.

The mechanism of MMLDE is as follows: after the initial surrogate model is constructed with the LHS samples, GP with the EI prescreening is used to determine the potential of the candidates for a certain number of iterations. The reason is that in this period the number of samples is not sufficient for constructing a reliable

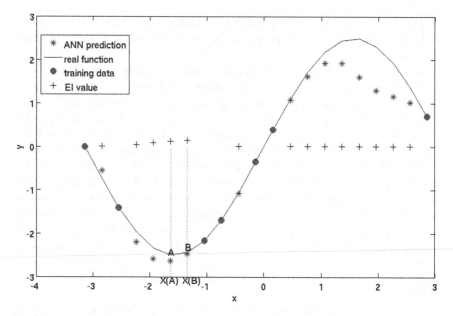

Fig. 8.8 An illustrative example for EI and ANN results in local search

surrogate model over the design space, so global search needs to be emphasized. When no improvement is obtained for a certain number of iterations using GP, the ANN prediction values are used to define the potential for improvement. If ANN provides no improvement for a certain number of iterations, GP and EI are activated again. This process is iteratively repeated until the termination condition is met.

8.4.3 The General Framework of MMLDE

The detailed flow diagram of MMLDE is shown in Fig. 8.9 and the description is in Fig. 8.10.

It is important to consider the cost of surrogate modeling. Chap. 7.5 has shown the computational complexity of the GP surrogate modeling. The number of training data affects the computational cost the most and the dimensionality is also an important factor. For larger than 20-dimensional problems, the GP model training cost is indeed a problem, as will be further discussed in Chap. 10. However, for most passive components, the number of design variables is often less than 5. For such dimensionality, both the GP model and the ANN model training costs from tenths of a second to a few seconds in our experiments, which can be neglected compared to the cost of the EM simulations.

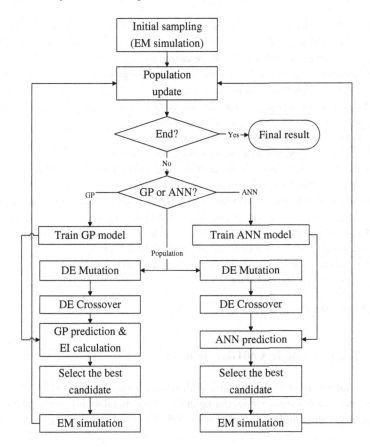

Fig. 8.9 Flow diagram of MMLDE

8.4.4 Experimental Results of MMLDE

In this section, the MMLDE algorithm is demonstrated for the synthesis of a 60 GHz inductor and a 60 GHz transformer in a 90 nm CMOS technology. The top two metal layers are used. ADS-Momentum [17] is used as the EM simulator. MMLDE stops when the performance cannot be improved for 40 consecutive iterations. The results shown below are based on 10 runs with independent random numbers. Time data reported in the experiments are wall-clock time. No parallel computation is used. EMGO is the reference method, which uses the same DE optimization but only EM simulations are used to evaluate performances. EMGO and other reference methods stop when the performance cannot be improved for 40 consecutive iterations.

Step 0: Initialize the parameters, e.g., threshold on the number of iterations to switch between GP and ANN, the DE algorithm parameters (e.g., CR), the GP parameters (e.g., the correlation function), the ANN parameters (e.g., the number of neurons, the training algorithm).

Step 1: Initialize the population by LHS sampling of the design space. The EM simulations are performed for the sampled design points.

Step 2: Check if the stopping criterion (e.g. a convergence criterion or a maximum number of iterations) is met. If yes, output the result; otherwise go to step 3.

Step 3: Judge to use the GP or ANN machine learning techniques.

Step 4: Train the selected surrogate model according to the available samples.

Step 5: Use the available samples as the current population, and perform the DE mutation operation (see Chapter 2) to obtain each candidate solution's mutant counterpart.

Step 6: Perform the DE crossover operation (see Chapter 2) between each candidate solution and its corresponding mutant counterpart to obtain each individual's trial individual.

Step 7: According to the selected model in Step 3, use the EI or the predicted value to select the individual with the possible best potential and perform the EM simulation.

Step 8: Update the population by adding the point from step 7 and its performance. Update the best solution obtained so far. Go back to Step 2.

Fig. 8.10 The MMLDE algorithm

8.4.4.1 Test Example 1: A 60 GHz Inductor

The first example is a 60 GHz inductor with circular shape in a 90 nm CMOS process. The design variables are the inner diameter (din), the metal width (mw) and the metal spacing (ms). The number of turns (nr) is 1.5. The synthesis problem is as follows:

$$\begin{aligned} &\text{maximize } Q \\ &\text{s.t. } L \in [0.45, 0.5] \text{ nH} \\ &SRF > 100 \text{ GHz} \end{aligned} \tag{8.7}$$

where Q is the quality factor, L is the inductance, and SRF is the self-resonance frequency.

The ranges of the design variables are:

$$\begin{aligned} din &\in [30, 100] \,\mu\text{m} \\ mw &\in [3, 10] \,\mu\text{m} \\ ms &\in [3, 8] \,\mu\text{m} \end{aligned} \tag{8.8}$$

Besides EMGO, SEMGO and the MMLDE framework working with either only the GP or only the ANN model are selected as reference methods. The latter two reference methods all use the same DE optimization engine and parameter settings as MMLDE, and are abbreviated as EMGOG and EMGOA, respectively.

The results are shown in Table 8.2 RCS is the number of designs satisfying the constraints over 10 different runs. N is the number of evaluations (average for the 10

Table 8.2 Results of different methods for example 1 (MMLDE)

Item	EMGO	EMGOG	EMGOA	SEMGO	MMLDE
RCS	10/10	9/10	2/10	5/10	10/10
Q	14.7	14.5	15.1	15.4	14.5
L	0.46	0.45	0.42	0.40	0.46
SRF	>100G	>100G	<100G	>100G	>100G
N	356	92	Not important (since specifications were not met) 96		40
T	19.1h	5.0h	Not important (since specifications were not met) 4.9h		2.1h

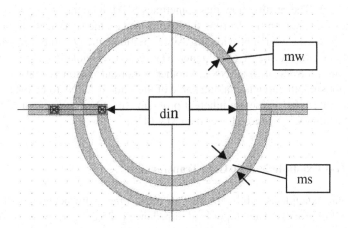

Fig. 8.11 A synthesized inductor for example 1

runs). T is the average time of 10 runs. The average performance for the optimization
goal Q and the typical performances of the constraints (L, SRF) are provided for each
method. For SRF, the corresponding Q at 100 GHz is calculated. If Q at 100 GHz is
larger than 0, the SRF is larger than 100 GHz.

It can be concluded from Table 8.2 that the result of MMLDE is comparable to that
of EMGO (the benchmark) but achieves an efficiency improvement of 10 times. Both
the design solutions of MMLDE and EMGO are near $[din, mw, ms] = [28, 5, 6]\,\mu m$.
The standard deviation of MMLDE on the optimization goal, Q, is 0.14. A typical
result of MMLDE for the 60 GHz inductor is shown in Fig. 8.11.

Next, we look at the performance of the MMLDE framework only with the GP
model and the EI prescreening (EMGOG), and only with the ANN model (EMGOA).
For the EMGOA method (column 4 in Table 8.2), we can see that the synthesis
does not succeed in many cases. This verifies that if only the predicted value is
used in on-line surrogate-model-based optimization, it is very easy to be trapped
in locally optimum points and the synthesis may fail. For the EMGOG method
we see that in most cases the synthesis is successful, because the EI prescreening
considers global exploration. However, it is about 2.5 times slower than MMLDE.
We found that the number of EM evaluations can show a large variation between

different runs using EMGOG: sometimes the necessary number of EM simulations is less than 30, but sometimes it can increase to more than 180 for a similar final result. This does not however occur in the MMLDE result because the switch of the two surrogate modeling and prediction methods improves the robustness. In case that one of them fails to find promising candidates, the other one will help. Hence, the hybrid surrogate modeling method in MMLDE enhances the speed and the optimization ability considerably. For the SEMGO method (column 5 in Table 8.2), different numbers of neurons and layers have been tested. Table 8.2 shows the results of 5 randomly selected ANNs from the group of trained ANNs whose training error is within 0.01. The results show that sometimes SEMGO obtains a good result and sometimes does not. In the 10 runs the results violate the specification on L for 5 times. In those runs which provide infeasible solutions, the Q value is much higher. Hence, the average value of Q is higher. The reason for violating the L constraint is the training errors of the ANNs. Another important reason is that the training data are not smooth enough, making some of the trained ANNs to be overfitted and lose generality [18]. These are inherent problems for ANNs. We can also find that SEMGO is almost 2.5 times slower than MMLDE.

8.4.4.2 Test Example 2: A 60 GHz Transformer

The second example is a 60 GHz overlay transformer with octagonal shape in a 90 nm CMOS process. The output load impedance is 25 $Omega$, which is the input resistance of the following stage. The design variables are the inner diameter of the primary inductor ($dinp$), the inner diameter of the secondary inductor ($dins$), the width of the primary inductor (wp) and the width of the secondary inductor (ws). The synthesis problem is as follows:

$$
\begin{aligned}
&\text{maximize } PTE \\
&\text{s.t. } k > 0.85 \\
&Q_1 > 10 \\
&Q_2 > 10 \\
&Re(Z_{in}) \in [10, 20]\Omega \\
&Im(Z_{in}) \in [10, 25]\Omega
\end{aligned}
\tag{8.9}
$$

where PTE is the power transfer efficiency, k is the coupling coefficient, Q_1 is the quality factor of the primary inductor, Q_2 is the quality factor of the secondary inductor, and Z_{in} is the input impedance at 60 GHz, which is the required optimal load impedance of the driver stage.
The ranges of the design variables are:

$$
\begin{aligned}
&dinp, dins \in [20, 150]\,\mu m \\
&wp, ws \in [5, 10]\,\mu m
\end{aligned}
\tag{8.10}
$$

Table 8.3   Results of different methods for example 2 (MMLDE)

Item		EMGO	GPDECO	MMLDE
RCS		10/10	10/10	10/10
PTE		89.0%	88.9%	88.8%
Average	*N*	965	207	87
Average	*time*	24.7 h	5.4 h	2.3 h

Fig. 8.12 A synthesized
transformer for example 2

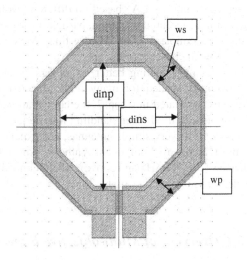

The comparison with GPDECO using the same EM simulator is carried out.[1] The transformer synthesis results are shown in Table 8.3. The average *PTE* for 10 runs is provided for each method. It can be seen in Table 8.3 that MMLDE costs 2.3 h on a single CPU node, which is very reasonable for practical use. Moreover, the result quality of MMLDE is comparable with the benchmark (the EMGO method), but MMLDE is more than 10 times faster. The standard deviation on the optimization goal, *PTE*, using MMLDE is 0.25%. A typical transformer result of MMLDE is shown in Fig. 8.12. Compared to the result of GPDECO, it can be seen that comparable results are obtained, but MMLDE is more than two times faster than GPDECO.

8.5 SAEA for Multi-objective Expensive Optimization and Generation Control Method

In the following sections, multi-objective RF/mm-wave passive component synthesis will be addressed. Some designers are interested in more than one performance and try to find their trade-offs. For instance, for integrated transformer design, RF

---

[1] in Sect. 8.2, a different EM simulator has been used, namely Sonnet, versus Momentum here.

designers may want to find a trade-off between efficiency and area. Multi-objective optimization is a useful method for these problems. Hence, multi-objective optimization algorithms have been increasingly applied to microwave engineering [19–22], covering microwave filters, antennas, microwave passive components and circuits in recent years.

In most approaches, multi-objective evolutionary algorithms (MOEAs) are selected as the search engine, due to their ability to approximate the Pareto front (PF). Several existing MOEAs based on different evolutionary algorithms (e.g., genetic algorithm, differential evolution, particle swarm optimization) are compared in [23] using the example of a microwave filter. Methods based on differential evolution (DE) show the best performance.

Besides investigating better optimizers, a more critical issue is the efficiency improvement needed for many multi-objective microwave component synthesis problems. As stated in Chap. 7, the computationally cheap equivalent circuit models can only be used when handling low-frequency (a few GHz) integrated passive components or very simple antennas, and expensive EM simulation is often a must for most microwave component synthesis, especially those working at mm-wave frequencies (e.g., 60 GHz and above), where lumped parasitic-aware equivalent circuit models are no longer accurate.

8.5.1 Overview of Multi-objective Expensive Optimization Methods

Compared to single-objective expensive optimization problems, multi-objective microwave component synthesis is much more difficult. The goal of multi-objective optimization is to generate the Pareto front, which includes a bunch of non-dominated points distributed in different areas of the design space, while only one of them is needed for single-objective optimization. In the CI field, the available prescreening-based surrogate model assisted multi-objective optimization algorithms, which are more efficient compared to prediction-based methods for small-scale problems, can obtain promising results for some benchmark test problems, but not for most of them [24, 25]. In the multi-objective microwave component synthesis area, the research on efficiency improvement is still an open topic. Most reported approaches directly use MOEAs as the search engine and EM simulators as the objective function evaluator [19, 20, 26]. Hence, the synthesis or optimization process is typically very CPU time expensive. Efficiency improvement based on hardware resources has been investigated in [20]. Because the evaluations of different candidate designs in a population are independent from each other in most MOEAs, parallel computation is used. However, to the best of our knowledge, there are very few efficient software algorithms for multi-objective microwave component synthesis. An example of multi-objective synthesis of an antenna using ParEGO [25] is reported in [27]. However, unlike SAEA for single-objective optimization, even state-of-the-art prescreening-based methods are found to be not so effective for multi-objective SAEA. For example, ParEGO [25] and MOEA/D-EGO [24], which are two state-of-the-art multi-objective prescreening-based SAEAs, are able to generate only parts of the Pareto front for many benchmark problems.

Multi-objective optimization requires a deeper understanding of the decision space compared to single-objective optimization, because the whole PF is generated, instead of a single solution. In the initial sampling and the first few iterations, the training data in many promising subregions may be far from sufficient. Hence, the constructed surrogate model often does not have an acceptable accuracy in those subregions, and the prescreening therefore cannot correctly locate new points in those promising areas for corresponding pieces of PF generation and surrogate model updating.

Another alternative is prediction-based methods. For small-scale problems, this may indeed cost more computational effort, but literature shows that high-quality results can be obtained [28]. For prediction-based methods, a reliable surrogate model is firstly constructed before using it (see Sect. 8.3.1). This is analogous to already having a good understanding and approximation of the hyper-surface. Then, multi-objective optimization is not more difficult than single-objective optimization when there are matured optimization methods. On the other hand, for prescreening-based method utilizing prediction uncertainty, many of the guesses may be wrong, because the guessing is much more complex than that of single-objective optimization.

8.5.2 The Generation Control Method

Generation control is widely used in prediction-based methods to avoid false optima. Because the surrogate model is constructed based on the available data, which are often not sufficient in the beginning of the optimization, the corresponding surrogate model may not be reliable, which needs to be taken into account in prediction-based methods. In other words, when using the current surrogate model to predict the newly generated candidate solutions without any control, the predicted values of some points may be far from their true objective function values. In this way, the search may converge to some false optimal points after several iterations [29], and the "good" candidates selected to perform EM simulations to update the surrogate model may also not be in the promising area. GPDECO makes use of the prediction uncertainty of GP modeling. For newly generated candidates with a large prediction uncertainty, EM simulation is used; for those with a small prediction uncertainty, the prediction value is used. However, the threshold value for the prediction uncertainty to judge if EM simulation should be used or not is based on the coarse model. Nevertheless, in many cases, there is not any coarse model available. In addition, using prescreening to generate a complete and high-quality Pareto front is still in its infancy. Therefore, a more general prediction-based method will be introduced, called the generation control method [30].

The basic idea of the generation control method is as follows. The total number of iterations (generations) is equally divided into several groups. For example, for 100 iterations, we can divide them into 1–10 iterations, 11–20 iterations,... 90–100 iterations. In each group, the population in some iterations uses expensive exact function evaluations for candidate evaluation, while for others prediction values are used. The number of iterations using expensive exact function evaluations in each group is adaptively adjusted based on the prediction uncertainty of the current surrogate model. Hence, when there are more available samples and the quality of the surrogate model is improved (as reflected by the prediction uncertainty), more iterations use prediction values, which enhances the efficiency considerably. On the other hand, when the quality of the surrogate model is not good enough, more expensive exact function evaluations are used to maintain correct convergence and to improve the surrogate model. This is the main idea of generation control for the ANN-based single-objective SAEA [30]. The rule how to adjust the number of iterations using expensive exact function evaluations in each group is as follows:

$$N_{EV}(k+1) = N_{EV(min)} + \lfloor \frac{S(k)}{S_{max}} \rfloor (N_{EV(max)} - N_{EV(min)}) \qquad (8.11)$$

where $N_{EV(max)}$ is the maximum number of iterations using expensive exact function evaluations in a group of iterations. $N_{EV(min)}$ is the minimum number of iterations using expensive exact function evaluations in a group. Note that in each group, at least one iteration needs to use expensive exact function evaluations in order to calibrate possible false optima. The $S(k)$ value is the maximum of all prediction uncertainties $\widehat{s}$ for the current population using the current surrogate model. S_{max} is the largest $\widehat{s}$ value found so far. Using the maximum $\widehat{s}$ value emphasizes the surrogate model quality, which is a safe setting.

8.6 Handling Multiple Objectives in SAEA

In Chap. 2, two widely used MOEAs were reviewed: NSGA-II and MOEA/D. When applying them together with surrogate models, each candidate solution has several predicted objective function values and corresponding prediction uncertainties. When prediction-based methods are used, the predicted objective functions values can be directly used and the problem is how to obtain a single value of prediction uncertainty for the use of the generation control method. It is straightforward to think of using the maximum, mean or minimum of the prediction uncertainties to describe the reliability of a prediction. They make a trade-off between the necessary number of expensive exact function evaluations and the risk of a false selection.

In MOEA/D, aggregation functions are used, such as Tchebycheff aggregation. The scalar function is (see Chap. 2):

$$\text{minimize} \quad g^{te}(x|\lambda, z^*) = max_{1 \leq i \leq m}\{\lambda_i | f_i(x) - z_i^* |\}, x \in \Omega \qquad (8.12)$$

where x is the vector of design parameters, $\lambda = (\lambda_1, \ldots, \lambda_m)$ is a weight vector and

$$\sum_{i=1}^{m} \lambda_i = 1 \tag{8.13}$$

$z^* = (z_1^*, \ldots, z_m^*)$ is the reference point in MOEA/D, and m is the number of objectives.

In MOEA/D evolution, different weight vectors will be used for each $f_i(x)$ to calculate the corresponding aggregation value. If the GP model is directly used to predict $g^{te}(x|\lambda, z^*)$, a large number of surrogate models will be generated, because for different λ and z^* (z^* is updated during the search) the objective function is different. [24] provides an approximation method to predict $g^{te}(x|\lambda, z^*)$ for different weight vectors and reference points by only using the surrogate models for the prediction of $f_i(x)$. Considering two objectives, their objective functions and aggregation functions are:

$$\begin{aligned} f_i(x) &\sim N(\widehat{y}_i(x), \widehat{s}_i^2(x)), i = 1, 2 \\ g_i^{te}(x) &\sim N(\widehat{y}_i^{te}(x), (\widehat{s}_i^{te}(x))^2), i = 1, 2 \end{aligned} \tag{8.14}$$

where $\widehat{y}_i^{te}(x)$ and $(\widehat{s}_i^{te}(x))^2$ are:

$$\widehat{y}_i^{te}(x) = \mu_1 \Phi(\alpha) + \mu_2 \Phi(-\alpha) + \tau \phi(\alpha) \tag{8.15}$$

$$(\widehat{s}_i^{te}(x))^2 = (\mu_1^2 + \sigma_1^2)\Phi(\alpha) + (\mu_2^2 + \sigma_2^2)\Phi(-\alpha) + (\mu_1 + \mu_2)\phi(\alpha) - (\widehat{y}_i^{te}(x))^2 \tag{8.16}$$

being

$$\begin{aligned} \sigma_i^2 &= [\lambda_i \widehat{s}_i(x)]^2 \\ \mu_i &= \lambda_i(\widehat{y}_i(x) - z_i^*), i = 1, 2 \\ \tau &= \sqrt{\sigma_1^2 + \sigma_2^2} \\ \alpha &= (\mu_1 - \mu_2)/\tau \end{aligned} \tag{8.17}$$

$\phi(\cdot)$ is the standard normal density function, and $\Phi(\cdot)$ is the standard normal distribution function.

8.6.1 The GPMOOG Method

A method, called Gaussian Process assisted multi-objective optimization with generation control (GPMOOG) [31], is presented for multi-objective synthesis of microwave components. The method aims to achieve comparable results as the traditional method (i.e., directly using MOEA with an EM simulator) for multi-objective EM-simulation-included synthesis, while highly improving the efficiency of the traditional method and making the computational time practical (from a couple of hours to about one day).

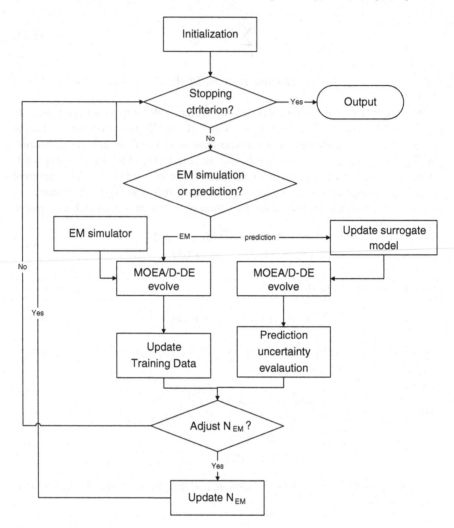

Fig. 8.13 The flow diagram of GPMOOG

MOEA/D-DE is used as the search engine. The flow diagram of GPMOOG is shown in Fig. 8.13. The GPMOOG algorithm works as follows[2]:

Input: (1) a multi-objective EM-simulation-included synthesis problem with m objectives

(2) a stopping criterion (e.g., maximum number of iterations)

---

[2] The algorithmic description for multi-objective optimization, to avoid the confusion of indices in a vector and the indices in a group of vectors, a superscript i indicates the ith individual of a group, and a subscript i indicates the ith element in a vector.

(3) MOEA/D parameters:

  N: the number of sub-problems;

  T: the neighborhood size;

  δ: the probability that parent solutions are selected from the neighborhood;

  n_r: the maximum number of solutions replaced by a child solution;

  λ: weight vector

(4) evolutionary search parameters:

  CR: crossover rate in DE;

  F: the scaling factor in DE

(5) Gaussian Process parameters: a correlation function

(6) Generation control parameters:

  C: the number of iterations in each group;

  $N_{EM(init)}$: the number of iterations using EM simulation in the first group;

  $N_{EM(min)}$, $N_{EM(max)}$ (see Sect. 8.5)

Output:

(1) approximated Pareto front;

(2) approximated Pareto set

Procedure:

Step 1: Initialization

Step 1.1: Compute the Euclidean distances between the weight vectors and work out the T closest weight vectors to each weight vector (the set is B). For $l = 1, \ldots, N$, set $B(l) = \{l_1, \ldots, l_T\}$. $\lambda^{l_1}, \ldots, \lambda^{l_T}$ are the T closest vectors to λ^l. Note that each sub-problem corresponds to a weight vector.

Step 1.2: Randomly generate an initial population $\{x^1, \ldots, x^N\}$. Calculate the fitness values of the population by EM simulations.

Step 1.3: Initialize $z = \{z_1, \ldots, z_m\}$, where $z_i = min_{1 \le l \le N} f_i(x^l)$.

Step 2: Update

For $l = 1, \ldots, N$,

Step 2.1: Selection of the mating pool:

Generate a random number which is uniformly distributed in [0,1]. Set

$$P = \begin{cases} B(l) & \text{if } rand < \delta \\ \{1, \ldots, N\} & \text{otherwise} \end{cases} \tag{8.18}$$

where $rand$ is random number in (0,1).

Step 2.2: Reproduction:

Set $r_1 = l$ and randomly select two indices r_2 and r_3 from P, and generate a new solution $\bar{y}$ by DE mutation (see Chap. 2). Then, perform a polynomial mutation (see Chap. 2) on $\bar{y}$ to produce a new solution y.

Step 2.3: Repair:

If an element of y is out of the bound of the design parameters defined beforehand, its value is reset to be a randomly selected value inside the boundary.

Step 2.4: Evaluation of the new candidate (y)

The current iteration is noted as $iter$.

(2.4.1) If $remainder(iter, C) < N_{EM}$ (N_{EM} is updated in the optimization process, see step 2.7) and $remainder(iter, C) \neq 0$, use EM simulation; otherwise, use GP surrogate model prediction.

(2.4.2) Evaluate the newly generated candidate y by the selected method. When using EM simulation, update the training data set by adding y and the corresponding performances from EM simulation. When using the GP surrogate model, update the S_{max} value.

Step 2.5: Update of the reference point:

For $i = 1, \ldots, m$, if $z_i > f_i(y)$, set $z_i = f_i(y)$.

Step 2.6: Replacement of solutions:

(2.6.1) For each l in P, calculate $g^{te}(y|\lambda^l, z)$ and $g^{te}(x^l|\lambda^l, z)$ by (8.15) and the $\hat{s}_i^{te}(y)$ value by (8.16).

(2.6.2) Set $c = 0$. If $g^{te}(y|\lambda^l, z) \leq g^{te}(x^l|\lambda^l, z)$, $c = c + 1$

(2.6.3) If $c \leq n_r$, for each l with $g^{te}(y|\lambda^l, z) \leq g^{te}(x^l|\lambda^l, z)$, set $x^l = y$. If $c > n_r$, for each l with $g^{te}(y|\lambda^l, z) \leq g^{te}(x^l|\lambda^l, z)$, calculate the Euclidean distances between $f(y)$ and $f(x^l)$ and then rank them. Choose n_r solutions with the smallest distances. Set $x^l = y$.

Step 2.7: Adjustment of N_{EM}:

If $remainder(iter, C) = 0$,

(2.7.1) Using all the $\hat{s}_i^{te}(y)$ from step (2.6.1), calculate the S value of the current surrogate model for the current population P by the method described in Sect. 8.5.

(2.7.2) Adjust N_{EM} using (8.11).

Step 2.8: Update iter:

$iter = iter + 1$

Step 3: Stopping Criterion:

If the stopping criterion is satisfied, then stop the algorithm and output $\{x^1, \ldots, x^N\}$ go to **Step 4**. Otherwise, go to **Step 2**.

Step 4: Output:

Perform EM simulation to the last population $\{x^1, \ldots, x^N\}$ to obtain their performances $\{f(x^1), \ldots, f(x^N)\}$. Delete dominated points and output the final PF.

8.6.2 Experimental Result

The GPMOOG algorithm is demonstrated for a 60 GHz transformer in a 90 nm CMOS technology. The parameters and their ranges are the same as the example in MMLDE, but the optimization problem is changed. The constraints of the quality factor, k factor and impedance matching are removed and the two objectives are the maximization of the power transfer efficiency (*PTE*) and the minimization of the square root of the area. ADS-Momentum is used as the EM simulator. For comparison, the traditional method (MOEA/D-DE with the EM simulator without surrogate models) has been run as a reference to compare both the generated PF and the efficiency. The generated

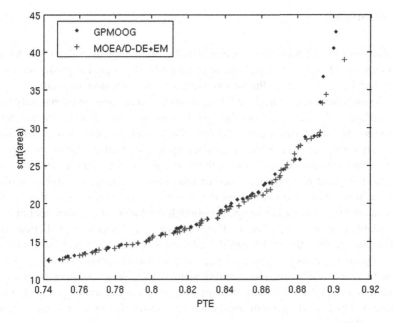

Fig. 8.14 PF generated by GPMOOG and the traditional method

PFs of both GPMOOG and the traditional method of directly applying MOEA/D-DE are shown in Fig. 8.14. Each point in the PF corresponds to a design.

From Fig. 8.14, it can be seen that the PF generated by directly using MOEA/D-DE with the EM simulator is only slightly better than that of GPMOOG. To quantify the difference, we use:

$$err(PF_{GPMOOG}, PF_{MOEA/D+EM}) - err(PF_{MOEA/D+EM}, PF_{GPMOOG})$$
$$(8.19)$$

where $err(A, B)$ is defined as: for each point a in A which is dominated by some points in B, we calculate $e = (\sum_{i=1}^{m} |a_i - \widehat{b_i}|/|\widehat{b_i}|)/m$, where m is the number of objectives, and $\widehat{b_i}$ is the nearest nondominated point to a_i in B. err is the average of the e values over all the points. The resulting PF generated by the traditional method is 2.88 % better than GPMOOG. This shows that the surrogate model in GPMOOG has a good performance. In the first two groups (1–5 iterations, 6–10 iterations), 4 and 3 iterations use EM simulations, respectively. After that, the number of iterations that use EM simulation falls to 1 in all the other groups. This confirms the adaptive adjustment of N_{EM}, and the final results verify the adjustment mechanism. The time consumption of GPMOOG (wall-clock time) is 28.2 h, which is reasonable for use in practice. In the 50 iterations, 16 of them use EM simulation. When directly using MOEA/D-DE with EM simulation, the time consumption is 3.6 days. Hence, a more than three times speed enhancement is achieved for about the same quality of result.

8.7 Summary

The fundamentals of surrogate model assisted evolutionary algorithm (SAEA) have been introduced in this chapter. Special attention has been paid to prediction uncertainty handling in SAEA. Prediction methods, prescreening methods and their hybridization have been introduced. Prediction-based methods are conservative and safe methods. The idea is to use the "good" surrogate model when the prediction uncertainty is lowered to a certain extent. They often consume more exact function evaluations but have a higher probability to avoid false optima compared to prescreening-based methods. Among them, the generation control method is general and has good performance. Prescreening-based methods are high reward but risky methods. The idea is to guess the possible promising areas of the decision space according to available samples, although a reliable surrogate model may not be available near the selected areas. They often perform better than prediction-based methods for small-scale single objective expensive optimization problems, while for multi-objective expensive optimization, prediction-based methods are often used. EI, PI and LCB are widely used prescreening methods, which can consider the quality of a candidate in a global picture by utilizing the prediction uncertainty, instead of minimizing it. The hybridization of them leads to considerably better results compared to any of them.

Three practical algorithms with different prediction uncertainty handling techniques for high-frequency integrated passive component synthesis have been discussed. The GPDECO and the MMLDE method are for single-objective constrained optimization of integrated passive components at high frequencies and the GPMOOG method focuses on the multi-objective optimization of integrated passive components. All of the three methods obtain comparable optimization quality to directly using computationally expensive EM simulations in an EA, and achieve a large speed enhancement.

However, we cannot stop thinking that if integrated passive components can be synthesized efficiently, why should it then not be possible to develop efficient global optimization methods based on the same idea, for the synthesis of high-frequency integrated circuits as well as complex antennas? This task is not trivial. The bottleneck is the "curse of dimensionality". In the following chapters, we will focus on this problem.

References

1. Nieuwoudt A, Massoud Y (2006) Variability-aware multilevel integrated spiral inductor synthesis. Comput Aided Des Integr Circ Syst, IEEE Trans 25(12):2613–2625
2. Mandal S, Goyel A, Gupta A (2009) Swarm optimization based on-chip inductor optimization. In: Proceedings of 4th international conference on computers and devices for, communication, pp 1–4
3. Niknejad A, Meyer R (2000) Design, simulation and applications of inductors and transformers for Si RF ICs. Springer, Boston

4. Mandal S, Sural S, Patra A (2008) ANN-and PSO-based synthesis of on-chip spiral inductors for RF ICs. Comput Aided Des Integr Circ Syst, IEEE Trans 27(1):188–192
5. Koziel S (2009) Surrogate-based optimization of microwave structures using space mapping and kriging.In: Proceedings of European Microwave Conference, pp 1062–1065
6. Bandler J, Cheng Q, Dakroury S, Mohamed A, Bakr M, Madsen K, Sondergaard J (2004) Space mapping: the state of the art. Microw Theory Tech, IEEE Trans 52(1):337–361
7. Liu B, He Y, Reynaert P, Gielen G (2011) Global optimization of integrated transformers for high frequency microwave circuits using a gaussian process based surrogate model. In: Proceedings of design, automation & test in Europe conference & exhibition, pp 1–6
8. Price K, Storn R, Lampinen J (2005) Differential evolution: a practical approach to global optimization. Springer-Verlag, New York.
9. Sonnet software homepage. http://www.sonnetsoftware.com/
10. Emmerich M, Giannakoglou K, Naujoks B (2006) Single-and multiobjective evolutionary optimization assisted by gaussian random field metamodels. Evol Comput, IEEE Trans 10(4):421–439
11. Jones D (2001) A taxonomy of global optimization methods based on response surfaces. J Global Optim 21(4):345–383
12. Jones D, Schonlau M, Welch W (1998) Efficient global optimization of expensive black-box functions. J Global Optim 13(4):455–492
13. Dennis J, Torczon V (1997) Managing approximation models in optimization: state-of-the-art, Multidisciplinary design optimization. IEEE Press, Panama
14. Lophaven S, Nielsen H, Søndergaard J (2002) DACE: a matlab kriging toolbox. Citeseer
15. Liu B, Zhao D, Reynaert P, Gielen G (2011) Synthesis of integrated passive components for high-frequency RF ICs based on evolutionary computation and machine learning techniques. Comput Aided Des Integr Circ Syst, IEEE Trans 30(10):1458–1468
16. Stein M (1987) Large sample properties of simulations using Latin hypercube sampling. Technometrics 29(2):143–151
17. Agilent Technology homepage. http://www.home.agilent.com/
18. Wasserman P (1989) Neural computing: theory and practice. Van Nostrand Reinhold Co, New York
19. Goudos S, Zaharis Z, Kampitaki D, Rekanos I, Hilas C (2009) Pareto optimal design of dual-band base station antenna arrays using multi-objective particle swarm optimization with fitness sharing. Magn, IEEE Trans 45(3):1522–1525
20. Poian M, Poles S, Bernasconi F, Leroux E, Steffé W, Zolesi M (2008) Multi-objective optimization for antenna design. In: Proceedings of IEEE international conference on microwaves, communications, antennas and electronic systems, pp 1–9
21. Oliveira D, Silva E, Santos J, Neto O (2011) Design of a microwave applicator for water sterilization using multiobjective optimization and phase control scheme. Magn, IEEE Trans 99:1242–1245
22. Brito L, De Carvalho P (2003) A general and robust method for multi-criteria design of microwave oscillators using an evolutionary strategy. In: Proceedings of the International Microwave and Optoelectronics Conference. IEEE, vol. 1, p:135–139
23. Goudos S, Siakavara K, Samaras T, Vafiadis E, Sahalos J (2011) Self-adaptive differential evolution applied to real-valued antenna and microwave design problems. Antennas Propag, IEEE Trans 99:1286–1298
24. Zhang Q, Liu W, Tsang E, Virginas B (2010) Expensive multiobjective optimization by MOEA/D with gaussian process model. Evol Comput, IEEE Trans 14(3):456–474
25. Knowles J (2006) ParEGO: a hybrid algorithm with on-line landscape approximation for expensive multiobjective optimization problems. Evol Comput, IEEE Trans 10(1):50–66
26. Chung K, Tam W (2008) Particle swarm optimization of wideband patch antennas. In: Proceedings of Asia-Pacific Microwave Conference, pp 1–4
27. John M, Ammann M (2009) Antenna optimization with a computationally efficient multiobjective evolutionary algorithm. Antennas Propag, IEEE Trans 57(1):260–263

28. Lim D, Jin Y, Ong Y, Sendhoff B (2010) Generalizing surrogate-assisted evolutionary compu-
    tation. Evol Comput, IEEE Trans 14(3):329–355
29. Jin Y (2005) A comprehensive survey of fitness approximation in evolutionary computation.
    Soft Comput Fusion Found, Methodologies Appl 9(1):3–12
30. Jin Y, Olhofer M, Sendhoff B (2002) A framework for evolutionary optimization with approx-
    imate fitness functions. Evol Comput, IEEE Trans 6(5):481–494
31. Liu B, Aliakbarian H, Radiom S, Vandenbosch G, Gielen G (2012) Efficient multi-objective
    synthesis for microwave components based on computational intelligence techniques. In: Pro-
    ceedings of the 49th annual design automation conference, pp 542–548

Chapter 9
mm-Wave Linear Amplifier Design Automation: A First Step to Complex Problems

In the previous chapter, the problem of the efficient synthesis of integrated passive components has been discussed. In this chapter, we will extend the design automation methods to mm-wave integrated circuits containing also active components. In Sects. 7.2 and 7.3, we have reviewed the current methods of RF integrated circuit design automation. Most of the existing RF IC design automation methods focus on low-GHz synthesis [1]. The usual approach is to use a (simple) lumped model to mimic the behavior of the key passive components (transformer, inductor). Regression methods are then used to fit the EM simulation results (S-parameters) to parasitic-included equivalent circuit models. Combining these with evolutionary algorithms and layout templates, the RF circuits can be synthesized. However, it has been discussed in Chap. 8 that equivalent circuit model-based methods are not applicable to mm-wave circuit synthesis. The reason is that at frequencies above 60 GHz, it is not accurate to represent a passive component by a simple lumped model over a wide bandwidth due to the distributed effects. However, research on building blocks from 40 to 120 GHz and beyond is increasing drastically and moving to industrial applications e.g., microwave imaging systems [2]. To keep the pace of the academic research and industrial applications, efficient synthesis methods for mm-wave amplifiers are needed.

The three SAEAs presented in Chap. 8 have the ability to solve small-scale expensive optimization problems, but considerable enhancements are needed to address the "curse of dimensionality". This chapter will present a solution method for linear mm-wave integrated circuit, called efficient machine learning-based differential evolution (EMLDE) algorithm [3]. This method aims to synthesize mm-wave circuits working at very-high frequencies (e.g., 100 GHz). For many mm-wave circuits based on CMOS technologies working at very high-frequencies, performances obtained by non-linear analysis, such as the efficiency, can hardly be optimized due to the limited f_T of the CMOS technologies. In other words, at these frequencies, "making the circuit work", or design for maximum gain, which is based on linear S-parameter analysis, is the main concern, and optimizing on other specifications (e.g., efficiency, which needs nonlinear harmonic balance simulation) is often very

B. Liu et al., *Automated Design of Analog and High-frequency Circuits*, Studies in Computational Intelligence 501, DOI: 10.1007/978-3-642-39162-0_9, © Springer-Verlag Berlin Heidelberg 2014

difficult. This problem represents a typical kind of problems in many engineering application areas:

(1) The problem has tens of decision variables (medium-scale).
(2) The problem is computationally expensive, but an evaluation of a candidate solution require several evaluation techniques, some of which are expensive while some are not.
(3) The performance of a candidate solution is given by all the decision variables, but the expensive evaluation only use a few of them (small-scale).

We can call such problems as decomposable expensive optimization problems.

The remainder of the chapter is organized as follows. Section 9.1 analyzes the main challenges, the key ideas and the structure of the presented mm-wave linear amplifier design automation system. Sections 9.2 and 9.3 discuss the key algorithms in the proposed system. Section 9.4 scales up the complete algorithm. The experimental verifications are provided in Sect. 9.5. Section 9.6 summarizes this chapter.

9.1 Problem Analysis and Key Ideas

9.1.1 Overview of EMLDE

EMLDE is an SAEA for medium-scale decomposable computationally expensive optimization problems. Compared to the SAEAs introduced in the evolution from passive components to complete circuits involve a higher dimensionality. Therefore, a hierarchical SAEA, called EMLDE, is designed. EMLDE aims to:

- develop a layout-included synthesis method for linear amplifiers working at mm-wave frequencies starting from a given circuit topology, specifications and some hints on layout (e.g., the metal layer to be used, the transistor layout template with different number of fingers);
- provide highly optimized results comparable to directly using an EA with EM simulations in the optimization loop, which is often the best known method with respect to the solution quality aspect;
- use much less computational effort than a synthesis method based on a standard EA, and as such make the computational time of the synthesis practical;
- be general enough for any technology and any frequency range in the mm-wave frequency band.

For the problem of RF amplifier synthesis, one stage of the amplifier often has 10–20 design variables and there are multiple stages. An example is provided in Sect. 9.5. However, most SAEAs concentrate on small-scale computationally expensive optimization problems (e.g., 5 dimensions) [4–6]. When the number of dimensions increases, some challenges appear. A linear increase of the number of design variables causes an exponential increase of the search space, which requires more training data in the on-line optimization process. This lowers the speed of the synthesis considerably, because more samples and iterations are needed and each evaluation is expensive. In addition, the computational effort to construct the GP model itself increases drastically (see Chap. 7 for details) with the number of training data and the number of design variables. When using prescreening-based methods, the correct detection of the possible promising areas of the design space is quite difficult due to the limited information (samples) from a large design space. When prediction-based methods are used, a large computational effort is indeed necessary to construct a reliable surrogate model because of the high dimensionality.

To address this problem, EMLDE focuses on dimension reduction to transform the original high-dimensional problem to a low-dimensional problem and on the efficient solution of this reformulated low-dimensional but more complex problem. The key ideas and main blocks in the flow will be described in the following subsections. The core algorithms will be described in a separate section.

9.1.2 The Active Components Library and the Look-up Table for Transmission Lines

In EMLDE, the active components library is the same as the parasitic-aware active components library used in existing low-frequency RF IC synthesis [1]. For transmission lines, a look-up table (LUT) is used to get the values of the S-parameters. Since the number of parameters of a transmission line is often small and the S-parameters of the transmission line change linearly with the design parameters, using a LUT is a reasonable choice to enhance the efficiency while maintaining the accuracy. In case of performance optimization of a linear RF amplifier using S-parameters (e.g., power gain), the computationally expensive parts are: the full parasitic extraction of active devices, the EM simulation of the long transmission lines and the EM simulation of transformers and inductors. When the performance optimization is based on linear analysis, usually the most critical problem is the impedance matching, instead of the transistor design. For example, the transistor length and a width fixed by the process are often used, while only the number of fingers is changed. In addition, the transistor layout is decided before any other components in many high-frequency amplifier designs. Hence, it is suggested to first extract the parasitics of the transistors with different number of fingers but with fixed width and length beforehand and directly use the extracted models in full-fletched optimization. The extraction consumes some computational effort, but it is a one-time investment for each technology. Note that changing the number of fingers brings a discrete design variable in the optimization. This is different from adjusting to grids for the other design parameters. Compared to the candidate values for other design variables (e.g., from a few μm to tens of μm), the rounding to grids (e.g., 1 nm) can almost be neglected, so we can still consider those variables as continuous variables. However, the number of fingers, which often needs to be rounded to the nearest integer, must be considered as an explicit discrete variable. A quantization technique [7] can be used to make the floating-point-based DE method also workable for mixed continuous and discrete optimization problems.

For the transmission lines, we first sweep the transmission lines with different line widths, lengths and distances between two lines when using differential transmission lines, and then build a LUT. This is a reasonably accurate and efficient method. Through experiments, it was found that the S-parameters generated by interpolation from the LUT differ little compared to the EM simulation results. Like the active components, the data generation for the transmission lines is also a one-time investment. This solves the efficiency problem of performing on-line EM simulations for newly generated long transmission lines. Therefore, the most expensive and difficult part remaining is the EM simulation of transformers and inductors, which cannot be solved by the above simple methods or equivalent circuit models when switching to mm-wave frequencies. Some of the S-parameters are very difficult to be trained and predicted with an acceptable error. Hence, the EM simulation of the inductors or transformers must be embedded in the optimization loop. This is also the focus of the efficient global optimization algorithm presented in this chapter.

9.1.3 Handling Cascaded Amplifiers

At high frequencies, the amplifier often cascades multiple stages to obtain a higher gain. For example, [8] reported the first fully differential 100 GHz CMOS amplifier, which uses 6 cascaded stages to obtain about 10 dB power gain. In manual design, the designer often copies the design of one stage to the other stages to construct the cascaded amplifier. This result is suboptimal because the required impedance matching of each stage is different. In contrast, the synthesis method proposed here optimizes the cascaded amplifier stage by stage according to each stage's own impedance matching. In EMLDE, transformers and inductors are the main objects for circuit division. The division rules are: (1) one stage includes one and only one computationally expensive passive component; (2) the components in each stage must be connected together; (3) there should not exist components that do not belong to any stage, such as the input/output pads. More details of the division will be shown in the example in Sect. 9.5.

9.1.4 The Two Optimization Loops

In this subsection, we introduce the method to reduce the number of dimensions of one stage of the RF amplifier. Usually, the design parameters of a stage of the RF amplifier include the parameters of the transformers or inductors, the parameters of the transistors and the parameters of the connecting transmission lines. The overall circuit performance is determined by all of them. But with the help of the active components library and the look-up table for transmission lines as discussed above,

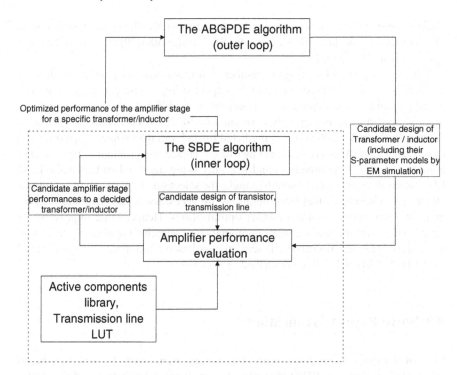

Fig. 9.1 The framework of the EMLDE method for the synthesis of mm-wave active circuits

only the parameters of the transformers or inductors need expensive EM simulation. In addition, the number of parameters of a transformer or inductor is not large (often 4–5). Hence, a natural idea is to separate these design variables. The method to reformulate the overall synthesis problem is shown in Fig. 9.1.

The parameters of the transformers or inductors are set as the design variables (input), and the performances of the amplifier with the candidate transformer or inductor and the corresponding *optimized* transistors and transmission lines are the output variables. In this way, the GP modeling can be used for the outer optimization loop to decrease the number of expensive EM simulations. In other words, the original plain optimization problem is reformulated as a hierarchical optimization problem. The outer loop is the optimization of the transformer or inductor parameters, whose function values are the optimized performances of the amplifier stage, which is obtained by the inner optimization loop. The inner loop is the optimization of the transistors and transmission lines for the decided transformer or inductor provided by the outer loop. Although the inner loop needs more computational effort, because an optimization is needed, rather than a single simulation, thanks to the efficient models for transistors and transmission lines, and also considering that the S-parameter simulation is computationally cheap, the evaluation of the inner function is cheap. In addition, because of the independence of the candidates in the population of EAs, parallel computation can be performed to further decrease the computational time. The selection-based differential evolution (SBDE) algorithm [9] is used for the inner optimization (see Chap. 2).

The price to pay for lowering the number of dimensions of the problem by decomposition is that the GP prediction and the expected improvement (EI) prescreening of the potential of a candidate design become more difficult. The reason is that the original performance is explicitly correlated to 10–20 or more variables, while in the new problem formulation it is predicted by 4–5 variables (outer loop) only and more than 10 variables are hidden (inner loop). Hence, the problem to be predicted is more complex. Experiments found that only using the standard GP method and EI prescreening is not good enough to make the selection of the promising solutions effectively. This means that more iterations are necessary, which naturally leads to more EM simulations and more inner optimizations. Hence, some improvements have been carried out that include a combination of adaptive population generation, naive Bayes classification and Gaussian Process-based differential evolution (called ABGPDE). ABGPDE will be described in Sect. 9.3.

9.2 Naive Bayes Classification

EMLDE includes decomposition into two optimization loops: the selection-based differential evolution (SBDE) algorithm for the inner optimization and the ABG-PDE algorithm for the outer optimization. The basics, Gaussian Process machine learning (see Chap. 7), expected improvement prescreening (see Chap. 8), differential evolution (see Chap. 2) and the selection-based differential evolution (see Chap. 2) have been discussed. Another basic technique to construct the key algorithm is naive Bayes classification. This section will introduce this technique.

As said above, by lowering the number of dimensions of the original optimization problem, the drawback is that the performances of the generated low-dimensional expensive optimization problem are more difficult to predict. Naive Bayes classification [10] is used to help the EI prescreening to select the most promising candidate design.

The naive Bayes classification is very efficient and outperforms many existing classification methods [11]. A classifier is a machine that maps the input feature space F to the output class label space C. Naive Bayes classification is a supervised learning method, which learns from a training data set of input vectors (input features) and their corresponding classes.

Assume that the input vector is d-dimensional, so we have feature variables from F_1 to F_d. Each input vector is classified to a class $C_i (i = 1, \ldots, n)$. For a new input vector x, the class it belongs to is decided by the maximum probability of the hypothesis that x belongs to C_i, that is:

$$class(x) = argmax_{C_i} Pr(C = C_i | F_1 = x_1, \ldots, F_d = x_d) \qquad (9.1)$$

The naive Bayes classifier assumes that each feature $F_k (k = 1, \ldots, d)$ is conditionally independent of every other feature. Hence, the conditional probability $Pr(C|F_1, \ldots, F_d)$ can be simplified to:

$$Pr(C|F_1, \ldots, F_d) = \frac{Pr(C, F_1, \ldots, F_d)}{Pr(F_1, \ldots, F_d)} = \frac{Pr(C) \prod_{k=1}^d Pr(F_k|C)}{Pr(F_1, \ldots, F_d)} \qquad (9.2)$$

where $Pr(F_1, \ldots, F_d)$ is common to the conditional probability of all classes and does not affect the ranking of Eq. (9.1). We assume that the input vector values associated with each class are Gaussian distributed. According to the training data, the mean and variance of the data associated with each class can be calculated. Using the probability density function of the Gaussian distribution and plugging Eq. (9.2) into Eq. (9.1), the corresponding class for a new vector x can be calculated.

Although the basic assumption of independence of all the features is often not accurate enough, the naive Bayes classifier uses the maximum of posteriori rule [10]. Therefore, the classification is decided by the ranking, rather than by the accurate estimation of $Pr(C = C_i|F_1 = x_1, \ldots, F_d = x_d)$. This is the main reason why the naive Bayes classifier can still be very effective, even when using such a simplified assumption.

9.3 Key Algorithms in EMLDE

9.3.1 The ABGPDE Algorithm

The ABGPDE algorithm is an SAEA for low-dimensional expensive optimization problems, especially suitable for problems with difficult to predict data sets. In EMLDE, the ABGPDE algorithm solves the outer loop optimization. The inputs are the design parameters of the transformer or inductor (4–5 dimensions), and the output is the performance of a stage of the linear amplifier with its corresponding optimized transistors and transmission lines. The function evaluation includes the EM simulation of the transformer or inductor and the entire inner optimization loop.

9.3.1.1 The Structure of ABGPDE

Due to the dimension reduction, the basic structure of MMLDE (see Chap. 8) can therefore be used in ABGPDE. The constraint handling method is the static penalty function method [12]. The differences with MMLDE are:

(1) The inner optimization is included. Both in initialization and optimization, an inner optimization is performed for each passive component design to obtain the corresponding optimal performance of the amplifier stage.

(2) Although the EI prescreening [4] is still used, the evaluation of the potential of the candidates is different from MMLDE.
(3) The population setting is different. In ABGPDE, there are two populations: one is the population containing all the simulated candidate designs and the other is adaptively constructed in each iteration as the parent population to generate child candidate designs.

9.3.1.2 Handling Difficult to Predict Data Sets

After lowering the original plain optimization problem with 10–20 design variables to a hierarchical optimization problem with 4–5 variables, the predictions are more difficult. Experiments show that the number of wrong selections using the EI pre-screening increases a lot.

Two techniques are used to address the problem: (1) improving the potential evaluation of the candidates; (2) revising the EA.

Using the EI prescreening, for a candidate whose tail of the probability density function is smaller than f_{min} (the smallest function value found so far, as discussed in Chap. 8), it can either truly be a promising one or a non-promising one with a large estimation variance. But EI cannot classify them. Hence, rather than improving EI, the naive Bayes classification is added to help EI for this classification. As said above, if the function is continuous, the function values of two points x_i and x_j should be close if they are highly correlated. The naive Bayes classifier, which only considers the input space x to make classifications, is used to help EI. This is a good supplement to GP machine learning which considers both the input space and the output space. If a candidate has a high EI value but is classified into the unpromising points class, there is a high probability that the point has an unpromising function value but with a large estimation variance.

In ABGPDE, the mean performance of the current circuit being synthesized is used for all the candidate designs of the current population as the threshold. The candidates with function value better than the threshold are classified as promising points; otherwise, they are classified as unpromising points. Both of them construct the training data. For a new population, the candidate solution with the highest EI value in the promising point class is considered as the most promising one and EM simulation is applied to it.

Although the naive Bayes classifier helps EI to evaluate the potential of the candidate solutions, it only contributes to the identification of promising solutions. On the other hand, the problem of how to make promising solutions be generated more efficiently still has to be discussed. The EA used for expensive optimization in MMLDE and EMLDE is different from the standard EAs. In standard EAs, the solution quality of the population is improving along the evolution process; so beneficial information to generate promising candidates keeps increasing in the consecutive populations. After some iterations, a high percentage of the information in the current population is beneficial to generate a candidate with good quality. In contrast, for prescreening-based surrogate model-assisted optimization, besides the initial samples, only one or few good new individuals are evaluated and added to the population in each iteration to increase the efficiency. Hence, the majority of the population correspond frequently the initial samples. The goal of the initial samples is to cover the design space but many of them may not be good solutions. Consequently, the percentage of beneficial information in the consecutive populations increases slowly. In evolutionary optimization, the new population is generated according to the information of the previous population by evolution operators. If the amount of beneficial information in the previous population is less, generating promising candidates is more difficult.

The idea to solve this problem is to artificially increase the amount of beneficial information by constructing a new population for evolution. The original population includes current simulated candidate designs, and the new population has the same size. In each iteration, all the simulated candidate designs are ranked and the top 75 % candidates are selected to enter the new population. The remaining vacancies of the new population are filled by randomly selecting the candidates from the top 75 % candidates of the original population. The new population is then used to generate child candidates. Faster convergence can be achieved using this technique.

Based on the above ideas, the ABGPDE algorithm is constructed. Experimental results show that ABGPDE enhances the speed and solution quality considerably compared to only using the GP-based surrogate model and EI prescreening [4].

9.3.2 The Embedded SBDE Algorithm

The SBDE algorithm is used for the inner optimization loop of the synthesis system (see Fig. 9.1). The inputs are the transistor parameters, transmission line parameters and DC voltages. The used transformer or inductor and its EM simulation result are provided from the outer loop. The output is the performance of one stage of the RF amplifier with optimized transistors, transmission lines and biasing voltages.

Although the S-parameter simulation for the circuit is fast, the inner optimization needs to evaluate a full population in each iteration and the optimization needs several iterations. Because the evaluation of different candidate designs in a population is independent from each other in SBDE, parallel computation is used. Experiments for the test circuit in Sect. 9.5 show that the inner optimization of a candidate transformer design can be completed in 5–6 min.

9.4 Scaling Up of the EMLDE Algorithm

Based on the above techniques, the scaling up of the EMLDE algorithm system is described in this section.

The general flow of the high-frequency linear RF amplifier synthesis system is shown in Fig. 9.1. The flow diagram of the EMLDE algorithm is shown in Fig. 9.2. ABGPDE and SBDE are included in EMLDE.

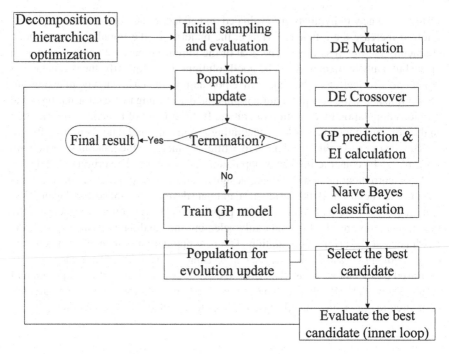

Fig. 9.2 Flow diagram of EMLDE

The EMLDE algorithm works as follows.

Step 0: Decompose the problem of optimizing an amplifier stage into a hierarchical optimization with outer and inner loops according to the method described in Sect. 9.1.

Step 1: Initialize the parameters, e.g., the DE algorithm parameters.

Step 2: Initialize the population by LHS sampling of the design space and perform EM simulation of the samples.[1] Perform the inner optimization loop for the samples. SBDE, the active component library and the transmission line LUT are used in this step.

Step 3: Update the population A by adding newly generated samples and their performances. In the first iteration, the added samples are from Step 2; afterwards, they are from Step 11. Update the best solution obtained so far.

Step 4: Check if the stopping criterion (e.g., a convergence criterion or a maximum number of iterations) is met. If yes, output the result; otherwise go to Step 5.

Step 5: Train the GP surrogate model according to population A.

Step 6: Construct the population for evolution (population B) as described in Sect. 9.3.

---

[1] This can be done beforehand for a given technology.

Step 7: Use population B and perform the DE mutation operation (see Chap. 2) to obtain each candidate solution's mutant counterpart.

Step 8: Perform the crossover operation between each candidate solution and its corresponding mutant counterpart (see Chap. 2) to obtain each individual's trial individual.

Step 9: Calculate the EI value of all the trial individuals from Step 8.

Step 10: Use the population A as the training data, apply naive Bayes classification as described in Sect. 9.2 to all the trial individuals from Step 8.

Step 11: Select the individual with the best potential according to the selection rules in the tournament selection-based constrained optimization method (see Chap. 2) and evaluate it using the same way as in Step 2. Go back to Step 3.

9.5 Experimental Results

9.5.1 Example Circuit

In this section, the EMLDE method is demonstrated for the synthesis of a 100 GHz three-stage transformer-coupled fully differential amplifier [8] in a 90 nm CMOS technology. One stage of the circuit configuration is shown in Fig. 9.3. Both transformers and transmission lines are used for impedance matching. Theoretically, a multi-turn transformer can achieve any impedance matching. However, the self-resonance frequency decreases as the number of turns of the transformer increases. Since this linear amplifier works at 100 GHz, a single-turn transformer is selected and transmission lines are used to improve the matching when needed. Using a very similar configuration but different sizing for each stage, the different stages are cascaded together. The synthesis problem is as follows.

$$\text{maximize power gain} @ 100\,\text{GHz}$$
$$\text{s.t. bandwidth} \geq 20\,\text{GHz} \tag{9.3}$$
$$k \text{ factors} > 1$$

Fig. 9.3 One stage of the 100 GHz amplifier (the full amplifier needs three such stages)

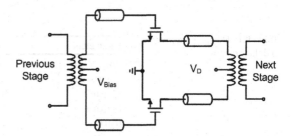

where the k factors are the Rollet stability factors [13]. Note that at 100 GHz, design for high gain is difficult due to the limitation of the transition frequency f_T of the 90 nm technology.

The transmission line used is a high-Q slow-wave coplanar transmission line [14]. The differential lines (CPW line) are on the top metal layer and the floating metal strips are on the lower metal layer. For the transformer, the top metal layer is used. All the transistors have the same size to ensure that each stage can drive the next stage. All the transistors have 1 μm width, 90 nm length and 15 fingers, as in [8]. In a cascaded multi-stage RF amplifier with the same transistor size, the optimal design parameters of the transformers often do not differ much from one stage to the other. As said above, the manual design method of copying the design of one stage to construct the whole amplifier achieves suboptimal results. Because of this, all the passive components are re-synthesized. But after performing the inner optimization on the initial samples in the synthesis of the previous stage, we can delete a few samples which have very bad performances, when synthesizing the next stage. This can avoid the inner optimization on these useless candidate points when synthesizing the next stage. The design variables are as follows. For transformers, the design variables are the inner diameter of the primary inductor ($dinp$), the inner diameter of the secondary inductor ($dins$), the width of the primary inductor (wp) and the width of the secondary inductor (ws) and the spacing between the two ports (sp). For transmission lines, the design variables are the metal width (lw), the metal length (ll), and the spacing between the differential lines (ls). Two DC voltages (V_D and V_{Bias}) are included in each stage. The input / output load impedance is 50 Ω. The ranges for the design variables are in Table 9.1, which are provided by the designer. There are 51 design variables for the three-stage amplifier. ADS-Momentum is used as EM simulator and HSPICE is used as the circuit simulator. Momentum supports parallel computation with multi-core computer and is used in EM simulations in all the methods.

The amplifier is synthesized stage by stage, starting from the output stage forward to the input stage. For the current stage being optimized, the passive components (transformer, transmission lines) are simulated separately, while the already synthesized stages are described by a single S-parameter model integrating all the passive components. For example, when optimizing stage 2, the transformer and the transmission lines have their own S-parameter models to enter the HSPICE simulation. For stage 3, which is already synthesized at that moment, the pad, transformer

Table 9.1 Design parameters and their ranges	Parameters	Lower bound	Upper bound
	$dinp, dins$(μm)	30	110
	wp, ws(μm)	2	10
	sp(μm)	8	23
	lw(μm)	1	10
	ll(μm)	2	80
	ls(μm)	7	23

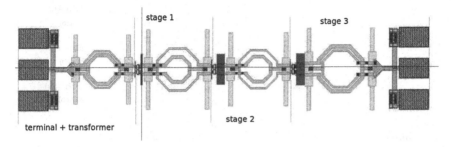

Fig. 9.4 The amplifier synthesized without machine learning

and transmission lines are connected together to perform an EM simulation, whose S-parameters result will be used when synthesizing stage 2. Because the amplifier includes four parts (see Fig. 9.4) to synthesize, and the matching impedances of each part are different, four parts are included in this example. The performance of the whole amplifier is affected by all of the four test problems, which tests the robustness of EMLDE.

9.5.2 Three-Stage Linear Amplifier Synthesis

The DE algorithm with EM simulation for the transformers without surrogate models is first applied. This method provides a high quality result, which can serve as reference result, but is of course very CPU time expensive. Parallel computation of the HSPICE simulations is not included. The synthesized circuit is shown in Fig. 9.4 and the simulation results are shown in Fig. 9.5. The power gain is 10.53 dB and the time consumption is about 9 days (wall-clock time).

Then, the EMLDE method is used. The synthesized circuit is shown in Fig. 9.6 and the simulation results are shown in Fig. 9.7. The power gain is 10.41 dB (i.e., slightly less than 10.53 dB) but the time cost is only 25 h (wall-clock time). The constraints are satisfied for both methods.

It can be seen that the EMLDE-based high-frequency RF linear amplifier synthesis method can achieve a comparable result compared to using the DE algorithm and EM simulation. We can also see the high solution quality from the mm-wave frequency RF IC design aspect. It is well known by designers that achieving 3 dB power gain per stage requires a very good matching at 100 GHz, and that the higher the working frequency, the more difficult to achieve high gain. The result shows that the average power gain of each stage reaches nearly 3.5 dB in the synthesized amplifier at 100 GHz which is very high considering the loss of the passive components and the pads. In terms of computational efficiency, about 9 times speed enhancement is achieved. The time cost of 25 h is very reasonable for practical use.

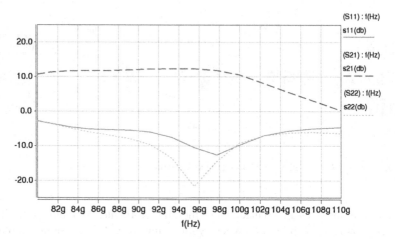

Fig. 9.5 S-parameters curve from 80 to 110 GHz (experiment 1)

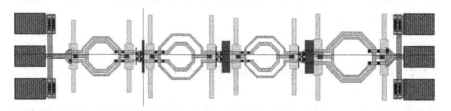

Fig. 9.6 The amplifier synthesized by EMLDE

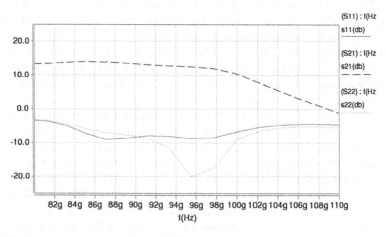

Fig. 9.7 S-parameter curve from 80 to 110 GHz (experiment 2)

In this synthesis, a total of 48 EM simulations are used (excluding the 49 initial sampling points which can be done beforehand in a given technology). The time spent on the EM simulations is 2.5 h. The inner loop optimization costs 22.7 h. When directly using the DE algorithm and EM simulation, nearly 4000 EM simulations are needed. It can be seen that EMLDE decreases the number of expensive EM simulations by about 80 times. Although more than 20 h are spent in the inner loop for circuit simulations in EMLDE, it is also much cheaper compared to many EM simulations. We can also conclude that the more complex the key passive components, which need more EM simulation time, the higher the advantage of EMLDE.

9.6 Summary

This chapter has introduced a practical algorithm, efficient machine learning-based differential evolution (EMLDE), for mm-wave frequency linear RF amplifiers. Only using the SAEA techniques introduced in Chap. 8, the process to establish an SAEA for medium-scale expensive optimization taking the advantage of the focused problem is shown. The core ideas are the decomposition method, which reformulates the problem into a hierarchical structure, and the proposed ABGPDE algorithm to solve the low-dimensional but more complex expensive optimization problems. From the experimental results, it can be seen that EMLDE provides comparable results to directly using a global optimization algorithm with EM simulations as performance evaluation, and achieves 9 times speed up. The objectives of achieving a high optimization ability, high efficiency and high generality are therefore met.

On the other hand, we also need to pay attention to the limitations of EMLDE. The foundation of the success of EMLDE is the decomposable structure of the mm-wave linear amplifier. Using this structure, the hierarchical method is proposed to reduce the dimensionality. This method however is less applicable when the decomposable structure does not exist or when the decomposition requires much effort, such as in the design automation of circuits working at relatively lower mm-wave frequencies or complex antenna synthesis. This touches upon the fundamental complexity limitation of SAEA. In the next chapter, we will address this problem by introducing state-of-the-art advanced innovations in SAEA research.

References

1. Allstot D, Choi K, Park J (2003) Parasitic-aware optimization of CMOS RF circuits. Springer, New York
2. Niknejad A, Meyer R (2000) Design, simulation and applications of inductors and transformers for Si RF ICs. Springer, New York, US
3. Liu B, Deferm N, Zhao D, Reynaert P, Gielen G (2012) An efficient high-frequency linear RF amplifier synthesis method based on evolutionary computation and machine learning techniques. IEEE Trans Comput Aided Des Integr Circuits Syst 31(7):981–993

4. Jones D, Schonlau M, Welch W (1998) Efficient global optimization of expensive black-box functions. J Global Optim 13(4):455–492
5. Jones D (2001) A taxonomy of global optimization methods based on response surfaces. J Global Optim 21(4):345–383
6. Zhang Q, Liu W, Tsang E, Virginas B (2010) Expensive multiobjective optimization by MOEA/D with gaussian process model. IEEE Trans Evol Comput 14(3):456–474
7. Price K, Storn R, Lampinen J (2005) Differential evolution: a practical approach to global optimization. Springer, New York
8. Deferm N, Reynaert P (2010) A 100 GHz transformer-coupled fully differential amplifier in 90 nm CMOS. In: Proceedings of IEEE Radio Frequency Integrated Circuits Symposium, 359–362
9. Zielinski K, Laur R (2006) Constrained single-objective optimization using differential evolution. In: Proceedings of IEEE Congress on Evolutionary Computation, pp 223–230
10. Larose D (2006) Data mining methods and models. Wiley, Hoboken
11. Caruana R, Niculescu-Mizil A (2006) An empirical comparison of supervised learning algorithms. In: Proceedings of the 23rd international conference on, machine learning, pp 161–168
12. Michalewicz Z (1995) A survey of constraint handling techniques in evolutionary computation methods. In: Proceedings of the 4th annual conference on evolutionary programming, pp 135–155
13. Rollett J (1962) Stability and power-gain invariants of linear twoports. IRE Trans Circuit Theory 9(1):29–32
14. Kaddour D, Issa H, Franc AL, Corrao N, Pistono E, Podevin F, Fournier J, Duchamp J, Ferrari P (2009) High-Q slow-wave coplanar transmission lines on 0.35 μm CMOS process. IEEE Microwave Wirel Compon Lett 19(9):542–544

Chapter 10
mm-Wave Nonlinear IC and Complex Antenna Synthesis: Handling High Dimensionality

In the previous chapter, the EMLDE method for efficient synthesis of mm-wave linear amplifiers has been presented. It is also a general solution for decomposable medium-scale computationally expensive global optimization problems. For these problems, the "curse of dimensionality" is addressed by the decomposition method. EMLDE transforms the medium-scale plain optimization problem to a hierarchical problem, in order to reduce the dimensionality. Nevertheless, decomposability is not possible for many real-world applications. For mm-wave ICs, EMLDE relies on the stage-by-stage design strategy for mm-wave linear amplifiers. Those amplifiers often work at quite high frequencies (e.g., 100, 120 GHz). At these frequencies, the power gain is often the main considered performance. Due to the limited f_T of many CMOS technologies, other performances, such as the efficiency, are often very low even if they are fully optimized. Thus, they are often not considered as optimization goal or specifications in synthesis. On the other hand, the optimization goal, the power gain, can be considered separately for each stage. In other words, the goal is to maximize the power gain in each stage and to combine the stages together. However, at relatively lower mm-wave frequencies (e.g., 30 to 60 GHz) for many technologies, a global optimization considering multiple performances is preferred if sufficient power gain is provided. A 60 GHz power amplifier is typically optimized for power added efficiency (PAE) and output power (the corresponding performance is 1 dB compression point, P_{1dB}) which requires harmonic balance simulations (nonlinear analysis). One stage designed for gain maximization may not be a good design for power added efficiency maximization, since the best impedance matchings for gain and efficiency are different. A possible solution is to apply appropriate constraints to each stage and then use EMLDE to synthesize the circuit stage by stage. However, determining appropriate performance specifications for each stage is often difficult even for experienced designers and failing to do so will lead to suboptimal designs. The other method is to consider the circuit as a whole. It is clear that the latter has advantages on achieving highly optimized results and may be useful to designers with any level of expertise. However, the decomposable structure is lost, introducing again the "curse of dimensionality". Furthermore, for complex antennas, it is almost

B. Liu et al., *Automated Design of Analog and High-frequency Circuits*, 201
Studies in Computational Intelligence 501, DOI: 10.1007/978-3-642-39162-0_10,
© Springer-Verlag Berlin Heidelberg 2014

impossible to decompose the synthesis task, and simultaneously considering the all design parameters of antenna may be the only possible solution. Therefore, general methods for medium-scale computationally expensive optimization problems are needed.

As has been discussed in Chap. 7, only mature SAEAs for low-dimensional problems (less than 10 dimensions) exist, and medium-scale SAEAs are an open research area. Typically, we call problems with 20–50 decision variables as medium-scale, and problems with 10–20 variables can also be effectively solved by SAEAs for medium-scale problems. This chapter will introduce state-of-the-art SAEA methods to handle higher dimensionality (i.e., 15–50 decision variables). Special focus will be paid to a method, called Gaussian Process surrogate model assisted evolutionary algorithm for medium-scale computationally expensive optimization problems (GPEME) [1], which well outperforms other methods for the same problems.

The remainder of the chapter is organized as follows. Section 10.1 discusses the main challenges brought by higher dimensionality and possible solution ideas. Section 10.2 introduces the general dimension reduction method based on Sammon mapping in GPEME. Section 10.3 introduces the surrogate model-aware search mechanism in GPEME. The experimental tests based on mathematical benchmark problems are in Sect. 10.4. In Sect. 10.5, GPEME is applied to a 60 GHz power amplifier. Three complex antenna synthesis examples are shown in Sect. 10.6. Finally, Sect. 10.7 summarizes this chapter.

10.1 Main Challenges for the Targeted Problem and Discussions

In computationally expensive optimization problems, the upper bound of exact function evaluations is determined by the practical (or user-imposed) time constraints. In simulation-based electromagnetic design automation, the computational time of a single simulation depends on the complexity of the structure and the requested accuracy. In many mm-wave frequency integrated circuit designs, an evaluation typically takes about 10 to 15 min. Due to very tight time-to-market requirements, the optimization should be finished within one or two weeks. Thus, one has to deliver the final solutions in about 1,000 function evaluations considering no parallel evaluation (e.g., using a cluster) is available. It is clear that the efficiency enhancement of such problem (or even more expensive problems) is very important, since without high-quality SAEAs, these problems are computationally intractable and cannot be solved.

When it comes to medium-scale problems (20–50 design variables), it is intuitive that more iterations are needed to find the globally optimal solution compared to small-scale problems. But there are more critical challenges than that. In Chap. 7, three cutting-edge methods for 20- and 30-dimensional expensive optimization problems [2–4] are reviewed. Reference [2] uses a weighted sum of the predicted values of different surrogate models to assist an EA, and the weights are adaptively adjusted based on the prediction uncertainty of different surrogate models. Reasonably good results for benchmark problems with 30 decision variables have been obtained, but it costs 8,000 exact function evaluations. Reference [4] uses the GP model with PI prescreening [5] as a global surrogate model and Lamarckian evolution (using the radial basis function) as a local surrogate model to accelerate an EA. Good results for 20-dimensional benchmark problems can be obtained with 6,000 exact function evaluations. [3] investigates GP-based local surrogate models and different prescreening methods to assist $\mu + \lambda$ evolution strategies (ES). Benchmark problems with 20 variables have been tested, and promising results have been obtained for some problems with only 1,000 exact function evaluations. However, its solution quality (especially for multimodal problems) needs to be improved.

An important challenge of the targeted problem (20–50 dimensional problems but only 1,000 allowed exact function evaluations) is the quality of the constructed surrogate model. The two main factors affecting the quality of the surrogate model are the number of training data and the mechanism of using the constructed surrogate model to assist the evolutionary search. It is intuitive that more training data are needed for medium- and high-dimensional problems to construct a reasonably good surrogate model: the higher the dimensionality, the more training data are necessary. On the aspect of the SAEA mechanism, when using the standard EAs as the search engine, like the above methods [2–4, 6], complex selections and replacements in various subregions require a globally reliable surrogate model. This can hardly to be achieved for the targeted problems since only 1,000 exact function evaluations are allowed at most.

In Chap. 7, the advantages of GP modeling have been discussed. An issue in an SAEA using GP modeling, particularly for problems with dozens of variables, is how to use a reasonable computational effort to build a good model for locating the most promising solutions. Note that GP modeling is not computationally cheap. The computational complexity of a typical learning algorithm for GP modeling can be $O(N_{it} K^3 d)$ [3], where N_{it} is the number of iterations for GP modeling, d is the number of variables (i.e., the dimensionality of the search space), and K is the number of training data. When d is large, a larger N_{it} is often necessary to obtain a good model. As a kind of reference, let us consider the time taken by the Blind DACE toolbox [7] on a Xeon 2.66 GHz computer in the MATLAB environment in the Linux system. The test problems are several benchmark problems (see the Appendix). If 100 training points are used, for 20-dimensional problems, the time to construct a GP model is about 10–20 s; for 30-dimensional problems, 15–35 seconds; for 50-dimensional problems, 120–250 s (2 to 3.5 min). When using 150 training data, constructing a 20-dimensional GP model costs 30–50 s; 30-dimensional, 80–110 s; 50-dimensional, 4 to 7 min. An SAEA with a budget of 1000 function evaluations may need to do GP modeling for several hundred times. Although using the optimized hyper-parameters from the GP model of the previous iteration as the starting point in a new GP model construction can reduce the computational overhead to some extent, it is not realistic to use too many training points in the GP modeling. On the other hand, using too few training points may deteriorate the reliability of the model and thus the solution quality.

There are two alternatives to address the challenges brought by higher dimensionality.

- The first idea is to introduce general dimension reduction techniques to address the "curse of dimensionality", especially for around 50-dimensional problems. When first transforming the decision space (the original space) with tens of variables to a lower-dimensional space with a few variables (the latent space) by dimension reduction techniques, we can then take advantage of Gaussian process model training in the low-dimensional space: the insufficient number of training data (with respect to the dimensionality) in higher dimensions becomes well sufficient for the lower-dimensional problem and the training cost is much lower. However, the transformation task is not trivial. The main problem is how to get a good mapping from the original space to the latent space with reduced size while maintaining the key information.

- The second idea is to replace the standard EA by a new search mechanism specially developed for SAEA. As was described above, most state-of-the-art SAEAs are based on standard genetic algorithms. This introduces complex population updating, which requires surrogate models with good quality in many subspaces to guarantee the correctness of replacements. Clearly, this is not good for surrogate modeling with limited training data. It becomes paradoxical in SAEAs that the evaluated candidate solutions are determined by the EA according to the optimization goals, but these solutions may not be the most appropriate ones for the surrogate modeling. This problem becomes more obvious when the training data is limited compared with the large design space, which occurs for the targeted problem. To that end, a surrogate model-aware evolutionary search mechanism which unifies the optimization and surrogate modeling is needed.

10.2 Dimension Reduction

10.2.1 Key Ideas

A key issue in the design of an SAEA can be stated as follows: Given K points $x^1, \ldots, x^K \in R^d$ and their fitness f-function values $y^1, \ldots, y^K$. Let $x^{K+1}, \ldots, x^{K+M}$ be M untested points in R^d. How can the M untested points without exact function evaluation be ranked?

A natural solution is to build a GP model based on the exactly evaluated y^i data, use a prescreening method for each untested point and then rank them according to their prescreened values. In GPEME, LCB prescreening is used (see Chap. 7) and only the point with the best LCB value is selected for exact function evaluation. It is desirable that the selected point is a high-ranked one among all the untested points in reality. Experiments have shown that for the test problems with more than 50 variables, it is often very hard to achieve the above goal with just 100 or 150 training data points. Using a larger K, as discussed in Sect. 10.1, can cause a large computational overhead in GP modeling.

To deal with problems with around 50 decision variables, an effective and efficient way is to use dimension reduction. The key idea is to transform the training data to a lower-dimensional space and then generate the model. As a result, the disadvantages caused by the insufficient number of training data for medium-scale problems are considerably relaxed for the resulting small-scale problems, since d is much smaller. The prescreening by GP with dimension reduction (GP + DR) method works as follows:

Input: (1) Training data: $x^1, \ldots, x^K$ and $y^1, \ldots, y^K$, (2) Points to prescreen: $x^{K+1}, \ldots, x^{K+M}$.

Output: The estimated best solution among all the untested points $x^{K+1}, \ldots, x^{K+M}$.

Step 1: Map $x^1, \ldots, x^{K+M} \in R^d$ into $\bar{x}^1, \ldots, \bar{x}^{K+M} \in R^l$, where $l < d$.

Step 2: Build a GP model by using the $\bar{x}^i$ and y^i data ($i = 1, \ldots, K$), and then use the model to compute the LCB value of each $\bar{x}^{K+j}$ ($j = 1, \ldots, M$). Output the x^{K+j} with the smallest LCB value as the estimated best solution.

There are many methods which can transform the original space R^d to the latent space R^l for dimensionality reduction [8]. Two factors are essential for selecting an appropriate dimension reduction method:

- The neighborhood relation among the data points plays a critical role in the correlation function in GP modeling (see Eq. (7.2)). Therefore, the neighborhood relationship among the $\bar{x}^i$ points in R^l should be as similar as possible to those among all the x^i points ($i = 1, \ldots, K + M$) in R^d. To that end, the pairwise distances among the data points should be preserved as much as possible.
- The mapping technique should not be very costly since it will be used many times in an SAEA.

The Sammon mapping [9] is a good choice which minimizes the differences between corresponding inter-point distances in the two spaces. A transformation is regarded as preferable if it conserves (to the greatest extent possible) the distances between each pair of points [10]. Moreover, it is not very expensive when $K + M$ and l are not large. Sammon mapping minimizes the following error function:

$$E = \sum_{1 \le i < j \le K+M} \frac{1}{dis(x^i, x^j)} \times \sum_{1 \le i < j \le K+M} \frac{[dis(x^i, x^j) - dis(\bar{x}^i, \bar{x}^j)]^2}{dis(x^i, x^j)} \quad (10.1)$$

where $dis(*, *)$ is the Euclidean distance.

Minimization of the error function in (10.1) is a large-scale quadratic optimization problem. Traditional mathematical programming methods can be applied, such as the gradient descent method [11]. For the initial point for the optimization process, it can be generated randomly or preferably by the principle component analysis (PCA) technique [10]. The gradient descent method in [11] with PCA technique to generate the starting point is used in GPEME. Note that the dimensionality of latent space cannot be large, so as to take advantage of the generated small-scale problem.

10.2.2 GP Modeling with Dimension Reduction Versus Direct GP Modeling

Sammon mapping transforms the training data points from the original space R^d to a lower-dimensional space R^l. Therefore, the number of training data points required for a good GP model can be reduced. In addition, since the modeling is conducted in a lower-dimensional space, the computational overhead can also significantly be reduced. However, although Sammon mapping minimizes the error function E between the two spaces, the error can hardly be zero. This means that some neighborhood information of the training data points in the original space will be lost in the latent space. The advantages and disadvantages of direct GP modeling in the original space R^d are exactly the opposite compared to GP modeling with dimension reduction (GP + DR).

Based on the above reasons, it is suggested to use the direct GP modeling when d is around 20 to 30, and GP + DR in the case of d is around 50.

10.3 The Surrogate Model-Aware Search Mechanism

This section introduces a new evolutionary search mechanism, called GPEME [1]. Like most other SAEAs, GPEME records all the evaluated solutions and their function values in a database. Once an exact function evaluation has been conducted for a new solution x, x and its real function value y will be added to the database. To initialize the database, a Design of Experiments (DoE) method, Latin Hypercube sampling (LHS) [12], is used to sample a set of initial points from the search space. This is done the same way as in MMLDE and has been widely used for initialization in some other SAEAs [13, 14]. Let the search space be $[a, b]^d$. The GPEME algorithm, whose flow diagram in shown in Fig. 10.1, works as follows:

Step 1: Use LHS to sample α solutions from $[a, b]^d$, evaluate their real function values and let them form the initial database.

Step 2: If a preset stopping criterion is met, output the best solution in the database; otherwise go to step 3.

Step 3: Select the λ best solutions (i.e., with the lowest function values) from the database to form a population P.

Step 4: Apply the DE operators on P to generate λ child solutions.

Step 5: Take the τ newest solutions in the database (i.e., the last τ solutions added to the database) and their function values as the training data to prescreen the λ child solutions generated in Step 4 by using GP in the original space or GP + DR with the LCB prescreening.

Step 6: Evaluate the real function value of the estimated best child solution from Step 5. Add this evaluated solution and its function value to the database. Go back to Step 2.

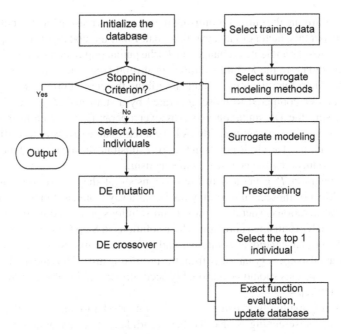

Fig. 10.1 The flow diagram of GPEME

Most SAEAs [2–4] use a single global surrogate model or a bunch of local surrogate models (e.g., a certain number of training data around each newly generated candidate solution) to predict the newly generated child population, some of which are then evaluated and used to update the current population and database in each iteration. An illustrative figure is shown in Fig. 10.2 for the prediction/prescreening of the cross point, which is in the promising area, as shown by the ellipse. Note that the locations of the training data (already evaluated candidate solutions) are largely determined according to the search mechanism of the optimization algorithm, rather than for producing high quality surrogate model(s).[1] The star points in Fig. 10.2 illustrates the locations of the training data pool in two of the d decision variables for a standard EA. Usually, the already evaluated candidate solutions spread in different search subregions because of the search operators in standard EAs. Let us assume τ' nearest points to the cross point are selected as training data. Experiments verify that when τ' is large, the points far away from the cross point will deteriorate the quality of the constructed surrogate model. To avoid this by decreasing τ', the training data are often not enough to produce a high quality surrogate model.

The key idea of the surrogate model-aware search mechanism is to improve the locations of the already evaluated candidate solutions to help surrogate modeling while keeping good optimization ability. The expectation is that the training data

---

[1] The opposite is the off-line surrogate modeling method, whose training data are purely for producing high quality surrogate model(s).

pool spreads around the current promising region and has reasonable diversity. There are two elementary factors to achieve this goal. First, the update of the population: Because at most 1 candidate is changed in P (the parent population) in each iteration, the best candidate in the λ child solutions in several consecutive iterations may be quite close. Therefore, the training data describing the current promising region can be much denser compared to those generated by a standard EA. Second, the DE scheme: Using the DE mutation and crossover operators, the λ child individuals generated from the λ parent individuals have a reasonable diversity to maintain the exploration ability. The circle points in Fig. 10.2 illustrate the locations of the training data pool produced by the new search mechanism.

The population P generated in Step 3 consists of the λ best solutions in the database. Most of these solutions may not be far away from each other, particularly after several iterations. Therefore, most child solutions generated in Step 4 are in a relatively small promising subregion. The τ solutions in Step 5 are the most recently generated optimal candidate solutions, and it is reasonable to assume that these solutions are also not very far away from the promising area. Thus the model built by using these solutions should be reasonably accurate for ranking the λ child solutions generated in Step 4.

Considering exploration, using a very accurate model to make sure that the best candidate by prescreening is also the best candidate in reality among the λ child solutions generated in Step 4 in each iteration is not necessary, or may even not be positive for the search. It is likely to happen that the top-ranked solutions in the child population are comparable in terms of optimality, and have reasonable diversity. Thus, it is satisfactory if the prescreened best solution in Step 5 is one of the top ranked solutions in reality, in terms of exact function values, among the λ child solutions.

Let us take the Ackley function with 20 variables as an example (see Appendix). GPEME is used to optimize the Ackley function with the experimental setting given in the next section. In the child population generated in each iteration, a prescreening-based best solution is selected, whose real ranking can be obtained by evaluating all the candidates in the child population. We found that over 20 runs, the average probabilities that the prescreening-based best solution is among the top 20 and 10 % of the λ child solutions are 86.1 and 75.8 %, respectively. Clearly, the GP modeling in GPEME works as expected.

To get a rough idea of the search ability of the new search framework, the GP modeling and prescreening are removed. Instead, exact function evaluations are used for the λ child solutions in each iteration, and randomly select one from the top β solutions. In such a way, we simulate the GP modeling and prescreening.

Setting $\alpha = 100$ and $\lambda = 50$, five different β values: 1, 3, 5, 7 and 9 are used. The function values of the best solution found so far versus the number of iterations is plotted in Fig. 10.3. These results are based on 20 runs. It can be seen that the convergence speed decreases as the β value increases. This indicates that the quality of the GP model does have an impact on the convergence speed of GPEME. When $1 \leq \beta \leq 7$, the best objective function value obtained with 1,000 iterations is very

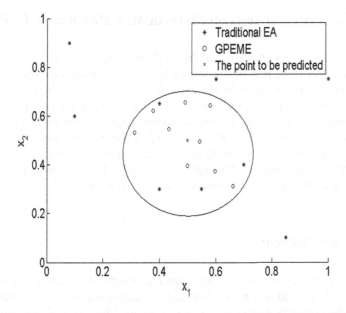

Fig. 10.2 An illustrative figure for the locations of the training data pool by different search mechanisms

Fig. 10.3 Convergence Curve in *simulated* GPEME

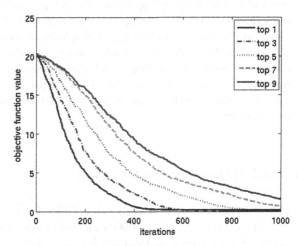

close to 0, the global optimal value. This implies that the surrogate model-aware search framework works well if the best solution determined by prescreening is among the real top-ranked candidates in the newly generated λ candidate solutions.

10.4 Experimental Tests on Mathematical Benchmark Problems

In this section, the GPEME algorithm will be demonstrated by mathematical benchmark problems. Benchmark test problems [15] are very important in intelligent optimization research. In the EC field, it is common to test and compare different algorithms using a set of benchmark functions. Benchmark functions are mathematical functions with different characters. Many of them are complex due to the properties of non-differential, multi-modal (more than one local optima), high dimensionality and non-separability (cannot be re-written as a sum of a set of functions with just one variable). Although they are difficult to optimize, their globally optimal solution(s) is/are often known, so as to evaluate an algorithm quantitatively.

10.4.1 Test Problems

The benchmark problems used in this chapter are selected from [2–4]. The test problems in [3, 4] have 20 dimensions. For single-objective unconstrained optimization, [3] uses the Sphere problem (simple unimodal), the Ellipsoid problem (unimodal), the Step problem (discontinuous) and the Ackley problem (multimodal). Reference [4] uses the Sphere problem, the Rosenbrock problem (unimodal but with a narrow valley near the global optimum), the Ackley problem, the Griewank problem (multimodal) and the Rastrigin problem (multimodal). Reference [2] uses test problems with 30-dimensional benchmark problems, including the Ackley problem, the Griewank problem, the Rosenbrock problem, two shifted rotated problems (more complex than the previous three benchmark problems) and five hybrid composition functions (very complex) from [16]. The problems used to test GPEME are listed in Table 10.1 and more details are in Appendix. 20, 30 and 50 dimensions are used in the test problems.

10.4.2 Performance and Analysis

F1 to F14 in Table 10.1 are used to test GPEME. 20 runs are performed for each problem. In all the problems, 1,000 exact function evaluations are used. Two results are presented for each problem in Figs. 10.4, 10.5, 10.6, 10.7, 10.8, 10.9, 10.10, 10.11, 10.12, 10.13, 10.14, 10.15, 10.16, 10.17. The first one is the convergence curve (part (a)), averaged over 20 runs. As described in Sect. 10.2, for the 20- and 30-dimensional problems, GPEME uses direct GP surrogate modeling in the original space without Sammon mapping. For the 50-dimensional problems, GPEME uses Sammon mapping and trains the GP model in the latent space. The opposite for 20/30- and 50-dimensional problems is used as comparison method. The scores of GPEME and the comparison method for problems F1 to F14 are shown in part (b) of

Table 10.1 The benchmark problems used to test the GPEME algorithm

Problem	Description	Dimension	Global optimum	Property	Reference Method
F1	Ellipsoid	20	0	unimodal	[3]
F2	Ellipsoid	30	0	unimodal	N.A.
F3	Ellipsoid	50	0	unimodal	N.A.
F4	Rosenbrock	20	0	unimodal with narrow valley	[4]
F5	Rosenbrock	30	0	unimodal with narrow valley	[2]
F6	Rosenbrock	50	0	unimodal with narrow valley	N.A.
F7	Ackley	20	0	multimodal	[3, 4]
F8	Ackley	30	0	multimodal	[2]
F9	Ackley	50	0	multimodal	N.A.
F10	Griewank	20	0	multimodal	[4]
F11	Griewank	30	0	multimodal	[2]
F12	Griewank	50	0	multimodal	N.A.
F13	Shifted Rotated Rastrigin	30	−330	complex multimodal	[2]
F14	Rotated Hybrid Composition Function (F19 in [2])	30	10	very complex multimodal	[2]

the Figs. 10.4, 10.5, 10.6, 10.7, 10.8, 10.9, 10.10, 10.11, 10.12, 10.13, 10.14, 10.15, 10.16, 10.17. Table 10.2 provides the statistics of the objective function values of GPEME over 20 runs. The computational time of GP model training in the original space in different dimensions has been discussed in Sect. 10.1. The computational time of the GP + DR method in the whole optimization process (900 times dimension reduction and low-dimensional GP model training) is typically 20–30 min. In GPEME, instead of using the predicted values to approximately replace the exact function evaluations, selecting the best candidate in the newly generated child population based on prediction and prescreening in each iteration is done. Because 100 exact evaluations are used for the initial surrogate model construction, the remaining 900 iterations are divided into 45 groups, and each group contains 20 iterations. In each iteration, there is a ranking and selection. It is interesting to investigate whether the real high-ranked candidates are selected by prescreening. For the benchmark problems, the real ranking of the selected "best" candidates in each iteration by exactly evaluating all the generated candidates is calculated.[2] The score of each subgroup of the 20 iterations is calculated as follows:

---

[2] This is only for test and is not included in GPEME.

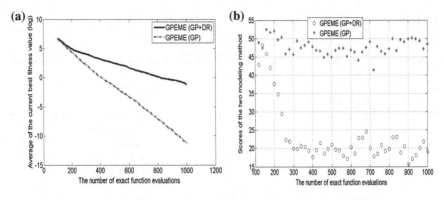

Fig. 10.4 Convergence trends and scores for benchmark problem F1 (20-D Ellipsoid)

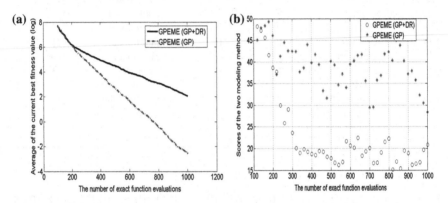

Fig. 10.5 Convergence trends and scores for benchmark problem F2 (30-D Ellipsoid)

$$Score = 3 \times size(N_{rank1}) + 2 \times size(N_{rank2}) + size(N_{rank3})$$
$$N_{rank1} = \{i\,|\,i \in top \quad 2\}$$
$$N_{rank2} = \{i\,|\,i \in top \quad 3 \quad to \quad 5\} \tag{10.2}$$
$$N_{rank3} = \{i\,|\,i \in top \quad 6 \quad to \quad 10\}$$

where "top n" is evaluated by exact function evaluations. $size(N_{rank1})$ refers to the number of prescreened best candidates which rank within top 2 in reality. $size(N_{rank2})$ and $size(N_{rank3})$ follow the same way but with different rankings. This score shows the quality of the surrogate modeling method for newly generated optimal candidates in GPEME. Indeed, if the selected candidate is out of the top 10 in reality, it cannot help the search much and therefore it is not counted in the score.

Some observations can be made from the above analysis. First, in most 20- and 30-dimensional problems, the benchmark problems are largely optimized by GPEME, and the optimized solutions are already near the global optimum in only 1,000 exact evaluations. The exceptions are the 20- and 30-dimensional Rosenbrock problem (F4 and F5), F13 and F14. Although the Rosenbrock problem is unimodal,

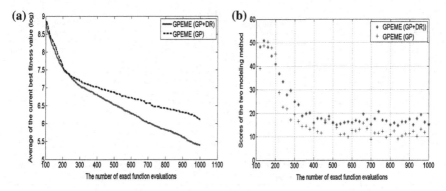

Fig. 10.6 Convergence trends and scores for benchmark problem F3 (50-D Ellipsoid)

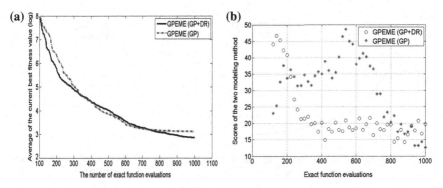

Fig. 10.7 Convergence trends and scores for benchmark problem F4 (20-D Rosenbrock)

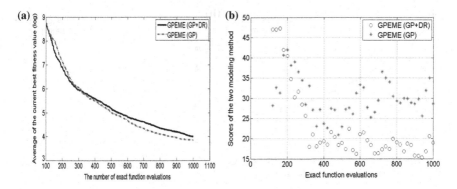

Fig. 10.8 Convergence trends and scores for benchmark problem F5 (30-D Rosenbrock)

the narrow valley of the global optimum makes it difficult to optimize. F13 and F14 are intentionally designed very complex problems. The standard DE is used with 30,000 exact function evaluations, and the obtained result are also not good enough for F5 (the 30-dimensional Rosenbrock problem), F13 and F14, which verifies the

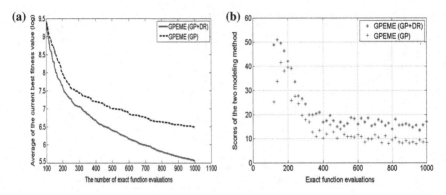

Fig. 10.9 Convergence trends and scores for benchmark problem F6 (50-D Rosenbrock)

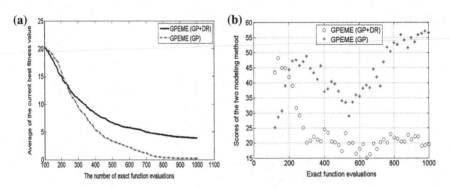

Fig. 10.10 Convergence trends and scores for benchmark problem F7 (20-D Ackley)

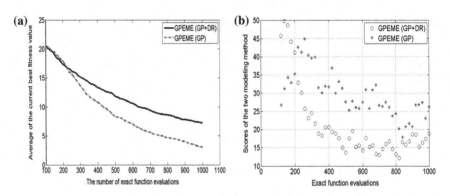

Fig. 10.11 Convergence trends and scores for benchmark problem F8 (30-D Ackley)

difficulty of these problems (see Table 10.3). For the 50-dimensional problems, it can be seen that the obtained results still have some distance from the global optimum. However, according to the corresponding convergence curves, the problems have been optimized to a large extent and the decreasing trend of the fitness function

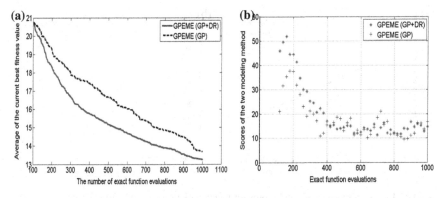

Fig. 10.12 Convergence trends and scores for benchmark problem F9 (50-D Ackley)

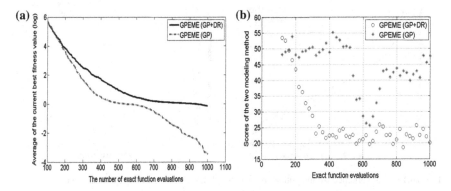

Fig. 10.13 Convergence trends and scores for benchmark problem F10 (20-D Griewank)

with the increase of the number of iterations is clear. Although only 1,000 exact evaluations can be used for the targeted problems, it can be expected that the fitness values will continue to decrease when more exact function evaluations are allowed. Because of the computational cost of constructing the GP model, the available hardware and requirement of many industrial problems, 100 training data are used for 50-dimensional problems. If more computational resources are available, more training data can be used and the result can also be improved.

It is necessary to compare the GP modeling in the original space and the GP modeling in the latent space constructed by Sammon mapping. From part (b) of Fig. 10.4 to Fig. 10.17, it can be seen that in most cases, the scores of the surrogate modeling methods are proportional to their fitness convergence. In other words, if more high-ranked candidates are correctly selected, the optimization will be more efficient. An exception is the F14 function. Figure 10.17 b shows that using Sammon mapping has better scores, but training the GP model in the original space gives a better fitness function value. However, F14 is an intentionally constructed hard problem and the

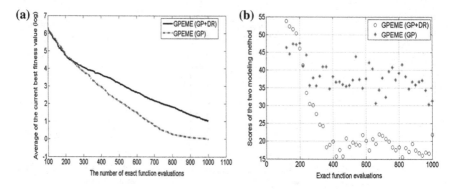

Fig. 10.14 Convergence trends and scores for benchmark problem F11 (30-D Griewank)

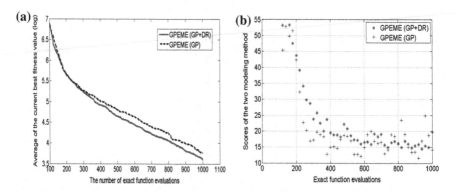

Fig. 10.15 Convergence trends and scores for benchmark problem F12 (50-D Griewank)

fitness function has not been optimized much according to Fig. 10.17a. Since this problem is also difficult to be optimized by most methods, such as traditional DE and GS-SOMA [2], the main observations are drawn from other figures.

From the convergence curves and the scores shown in Fig. 10.4 to Fig. 10.17, it can be seen that for most 20- and 30-dimensional problems, both using GP in the original space and in the latent space, can obtain an acceptable result. However, for 20 and 30 dimensions, using GP in the original space gets better results in most cases. Note that the performance of surrogate modeling methods is also affected by the properties of the function. For example, for the Ellipsoid problem, which is a simple unimodal problem, directly using GP in the original space highly outperforms the method of using the latent space for 20 and 30 dimensions (see Fig. 10.4b and Fig. 10.5b). On the other hand, for the 30-dimensional rotated shifted Rastrigin problem, whose surface is highly multimodal, using GP in the latent space is better since in the original space more training data (with respect to the dimensionality) are needed to approximate the surface (see Fig. 10.16b). In contrast, for the 50-dimensional problems, in all the tested problems, using Sammon mapping and building the GP model in the latent space (the GP + DR method) is better than constructing the GP model in the original

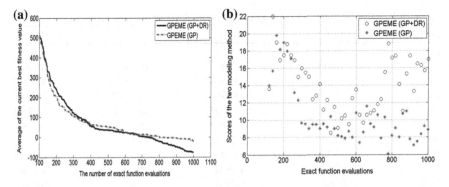

Fig. 10.16 Convergence trends and scores for benchmark problem F13 (30-D Shifted Rotated Rastrigin)

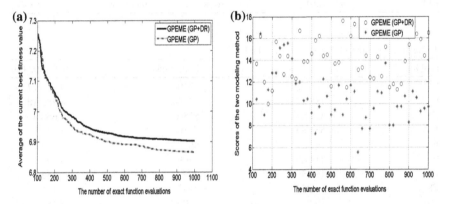

Fig. 10.17 Convergence trends and scores for benchmark problem F14 (30-D Rotated Hybrid Composition Function, F19 in [2])

space, even for the simple Ellipsoid problem. Therefore, we can conclude that using GP + DR for around 50-dimensional problems and using the GP model in the original space for around 20- and 30-dimensional problems is a reasonable empirical rule.

The standard DE [17] without surrogate model is used to optimize the test problems. The values of the population size, the scaling factor F and the crossover rate CR are the same as those used in GPEME. The DE/best/1 mutation operator is used in order to increase the efficiency. The number of exact function evaluations is 30,000. The results are shown in Table 10.3. In Table 10.3, "DE(1,000)" means the average value in 20 runs when 1,000 exact function evaluations are made. "DE(number)" means the number of exact function evaluations needed to achieve a similar result compared to GPEME which uses 1,000 exact function evaluations. "DE(30,000)" is the final result after 30,000 exact evaluations, which is used to indicate the difficulty of the problem.

Table 10.2 Statistics of the GPEME results in terms of objective function values for F1–F14

Problem	Best	Worst	Average	Std
F1	1.27e-6	7.2e-5	1.3e-5	2.18e-5
F2	0.0155	0.1647	0.0762	0.0401
F3	134.0681	372.5567	221.0774	81.6123
F4	15.1491	75.8806	22.4287	18.7946
F5	26.2624	88.2325	46.1773	25.5199
F6	172.3547	401.4187	258.2787	80.1877
F7	0.0037	1.8403	0.1990	0.5771
F8	1.9491	4.9640	3.0105	0.9250
F9	9.2524	14.9343	13.2327	1.5846
F10	0.0002	0.2234	0.0307	0.0682
F11	0.7368	1.0761	0.9969	0.1080
F12	22.5456	64.9767	36.6459	13.1755
F13	−57.0678	18.0327	−21.8610	36.4492
F14	933.1601	992.8618	958.5939	25.6946

Table 10.3 Standard DE results for F1–F14

Problem	DE (1,000)	DE (number)	DE (30,000)
F1	203.22	12400	9.47e-17
F2	788.05	10750	4.28e-10
F3	3281.3	5750	4.65e-4
F4	534.85	4550	1.24
F5	1621.1	5650	14.53
F6	4147.9	5800	43.70
F7	16.94	9000	4.18e-9
F8	18.35	8050	0.13
F9	19.24	4100	1.18
F10	81.06	11400	0.013
F11	183.18	9500	0.0025
F12	467.77	5400	0.0069
F13	184.09	3150	−138.15
F14	995.58	1600	918.84

From Table 10.3, the speed enhancement of GPEME can clearly be seen. Note that the implemented DE also uses the DE/best/1 mutation, so the speed enhancement of GPEME really comes from the surrogate modeling method and the surrogate model-aware search mechanism.

To verify the effectiveness of Sammon mapping, some other dimension reduction methods are used to replace Sammon mapping for the latent space construction in GPEME. F7 is used as an example. Four other dimension reduction methods are PCA, linear discriminant analysis (LDA), local linear embedding (LLE) and

Table 10.4 Result of F7 using five dimension reduction methods

Sammon	PCA	LDA	LLE	NCA
3.57	14.93	12.5	18.73	13.29

neighbourhood components analysis (NCA) [8]. The results based on 10 runs are shown in Table 10.4. It can be seen that Sammon mapping has clear advantages.

GPEME has been compared with several state-of-the-art SAEAs [2–4] and has shown clear advantages. For details, please refer to [1].

10.5 60 GHz Power Amplifier Synthesis by GPEME

In this section, a 60 GHz cascode power amplifier in a 65 nm CMOS technology is synthesized by GPEME. The circuit configuration is shown in Fig. 10.18. The PA has a driver stage and an output stage. The pseudo-differential cascode amplifier is used to increase the output voltage swing and improve the stability. Three experiments are done. The output stage has 6 transistors in the differential pair and in the first two experiments, the driver stage has 3 transistors in the differential pair, which is decided by the designer. In the third experiment, the number of transistors in the driver stage is set as a design variable, with a range of 2 to 4. This example represents a different kind of typical mm-wave circuit synthesis problem compared to Chap. 9. Different from PA working at very high frequency, for mm-wave ICs working at relatively low mm-wave frequencies (e.g., 40, 60 GHz), more performances need to be taken into account to meet the specifications of the standards [18, 19]. Taking the 60 GHz power amplifier for example, the output power, the efficiency and the power gain are important performances, which need to be considered together. Nevertheless, the design of the matching network is different for maximizing the efficiency or maximizing the gain. Therefore, the stage-by-stage synthesis method of EMLDE can hardly be used for this case. When the stage-by-stage synthesis method is applied, the designer has to provide appropriate specifications for each stage, and the difficulty of the trade-off between the power gain and the efficiency is left to the designer himself/herself. Hence, considering the circuit as a whole in the synthesis is better and more general. Therefore, for this kind of problems, GPEME is more appropriate.

Like EMLDE, the transistors layouts with different fingers are prepared beforehand by the designer. The transformers are implemented in an overlay structure using the top two metals layers. The design variables are the inner diameter of the primary inductor ($dinp$), the inner diameter of the secondary inductor ($dins$), the width of the primary inductor (wp) and the width of the secondary inductor (ws). There are 5 biasing voltages: V_{DD}, V_{cas1}, V_{cas2}, V_{b1} and V_{b2} (see Fig. 10.18). The ranges for the design variables are summarized in Table 10.5, and are determined by the designer. The output load impedance is 50 Ω. There are in total 17 design variables, which need to be optimized altogether.

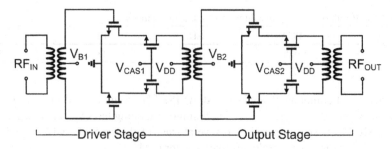

Fig. 10.18 Schematic of the 60 GHz power amplifier

Table 10.5 Design parameters and their ranges for the 60 GHz power amplifier

Parameters	Lower Bound	Upper Bound
$dinp, dins$ (μm)	20	100
wp, ws (μm)	3	10
V_{DD}(V)	1.5	2
V_{cas1}(V)	1.2	2
V_{cas2}(V)	1.2	2
V_{b1}(V)	0.55	0.95
V_{b2}(V)	0.55	0.95

Two optimization problems with different objective and constraints are defined. The first one optimizes the 1 dB compression point with constraints on the power added efficiency and the power gain, while the second one optimizes the power added efficiency with constraints on the 1 dB compression point and the power gain. In addition, we will consider the power added efficiency at P_{sat} and P_{1dB}. The former one is often used as a performance of a power amplifier in the literature, while the latter one is more important in practical design. The first optimization problem is:

$$\text{maximize 1 dB compression point}$$
$$\text{s.t. power added efficiency (at } P_{sat}) \geq 12\,\% \tag{10.3}$$
$$\text{power gain} \geq 15\,\text{dB}$$

ADS Momentum is used as EM simulator to evaluate the S-parameter models of the transformers. Cadence SpectreRF is used as the RF simulator and harmonic balance simulations are performed to obtain the power added efficiency, the 1 dB compression point and the power gain for a candidate design (i.e., a 60 GHz power amplifier with parasitic-extracted transistor models and S-parameter models of the transformers).

The layout of the synthesized power amplifier is shown in Fig. 10.19. The 1 dB compression point is 12.42 dBm, the power added efficiency at P_{sat} is 12.00 % and the power gain is 15.73 dB. The simulation results are shown in Fig. 10.20.

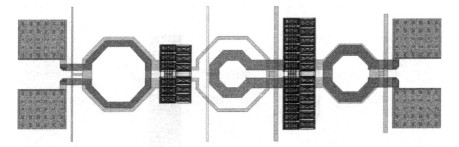

Fig. 10.19 Layout of the power amplifier synthesized by GPEME (problem 1)

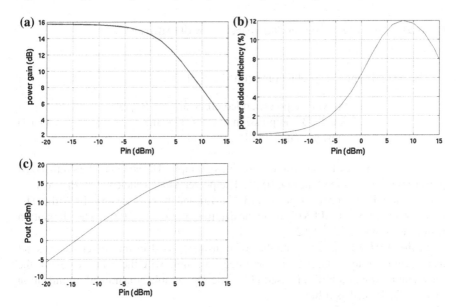

Fig. 10.20 The simulated performances of the 60 GHz power amplifier (problem 1) synthesized by GPEME. **a** Simulated power gain of the 60GHz power amplifier (problem 1). **b** Simulated power added efficiency of the 60GHz power amplifier (problem 1). **c** Simulated output power of the 60GHz power amplifier (problem 1)

The time consumption of GPEME to synthesize this power amplifier is about 2 days (wall-clock time). To derive the speed enhancement compared with the EMGO method (see Chap. 10) may cost a too long time. Therefore, the upper limit of the synthesis time using the EMGO method is set to 10 days. After 10 days, the results obtained by DE with the tournament selection-based method (i.e., SBDE) are as follows. The 1dB compression point is 9.97 dBm, the power added efficiency at P_{sat} is 8.48 % and the power gain is 12.16 dB. It can be seen that the result obtained by SBDE in 10 days is far from optimal. In contrast, GPEME obtains a highly optimized result in only 51 h.

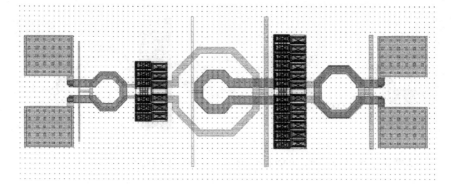

Fig. 10.21 Layout of the power amplifier synthesized by GPEME (problem 2)

The second synthesis problem is defined as follows:

$$\text{maximize power added efficiency (at } P_{1\,dB})$$
$$\text{s.t. 1 dB compression point} \geq 12\,dBm \qquad (10.4)$$
$$\text{power gain} \geq 14\,dB$$

All the settings are the same as the previous example. The layout of the synthesized power amplifier is shown in Fig. 10.21. The power added efficiency at P_{1dB} is now 6.47 %, the 1 dB compression point is 13.45 dBm, and the power gain is 14.00 dB. The time consumption of GPEME to synthesize this problem is 50.6 h. The simulation results are shown in Fig. 10.22.

In the third example, the number of transistors used in the driver stage is set as a design variable. Because it is an integer variable with the values of 2/3/4, the quantization method in DE is used [17]. The 18-variable constrained optimization problem is defined as follows.

$$\text{maximize power added efficiency (at } P_{1dB})$$
$$\text{s.t. 1 dB compression point} \geq 12\,dBm \qquad (10.5)$$
$$\text{power gain} \geq 15\,dB$$

The layout of the synthesized power amplifier is shown in Fig. 10.23. The power added efficiency at P_{1dB} is 5.32 %, the 1dB compression point is 12.61 dBm, and the power gain is 15.24 dB. The optimized number of transistors for the driver stage is 4. The time consumption of GPEME to synthesize this problem is 52.5 h. The simulation results are shown in Fig. 10.24. In all the examples, the Rollet stability factors (K factors) are larger than 1, which guarantee the stability of the PA. For example, in the third experiment, the minimum value of the K factors is around 50 GHz, and is about 5.5. Therefore, the synthesized PAs are unconditionally stable.

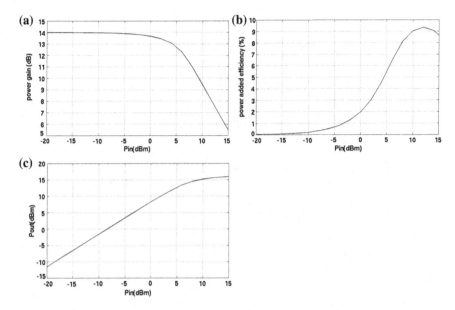

Fig. 10.22 The simulated performances of the 60 GHz power amplifier (problem 2) synthesized by GPEME. **a** Simulated power gain of the 60 GHz power amplifier (problem 2). **b** Simulated power added efficiency of the 60 GHz power amplifier (problem 2). **c** Simulated output power of the 60 GHz power amplifier (problem 2)

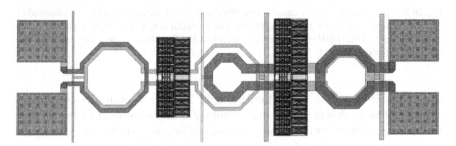

Fig. 10.23 Layout of the power amplifier synthesized by GPEME (problem 3)

10.6 Complex Antenna Synthesis with GPEME

In this section, GPEME is used for complex antenna synthesis. Antenna optimization problems are usually very complex due to the large number of variables and the sensitivity of the desired goals respect to these variables. Three antenna examples are selected: a 1.6 GHz microstrip-fed crooked cross slot antenna (17 design parameters), a 60 GHz on-chip antenna for wireless communication (10 design parameters) and a four-element linear array antenna (19 design parameters). ADS-Momentum [20] is used as the EM simulator for the first two examples and Magmas [21–23] is

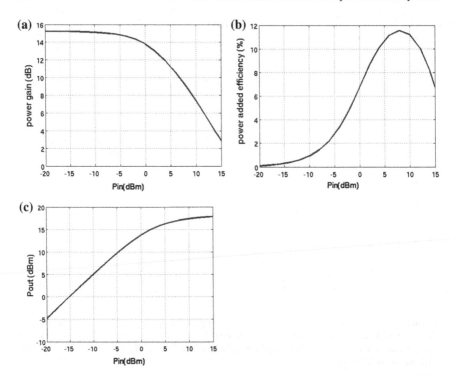

Fig. 10.24 The simulated performances of the 60 GHz power amplifier (problem 3) synthesized by GPEME. **a** Simulated power gain of the 60GHz power amplifier (problem 3). **b** Simulated power added efficiency of the 60GHz power amplifier (problem 3). **c** Simulated output power of the 60GHz power amplifier (problem 3)

used as the EM simulator for the third example. The bounds of the design variables are set both by the design rules of the technology and the experience of the designer. Like all the problems in this book, the ranges of design variables provided by the designers are quite large, and not much experience is needed. For the first two examples, GPEME stops when the performance cannot be improved for 30 consecutive iterations. For the third example, since it is computationally quite expensive, the number of EM simulations is restricted to 600. All the time consumptions reported in the experiments are wall-clock time. To evaluate the quality of the solution provided by GPEME, the reference method we have selected for the first two examples is the selection-based differential evolution (SBDE) algorithm for single-objective constrained optimization (see Chap. 2). For the third example, we compared with PSO with the same settings in [24] but using the fitness function from Sect. 10.6.3, which is better than the published result in [24] according to Sect. 10.6.3. Obviously, the reference methods are CPU time expensive, but often can provide the best result as reference to test GPEME.

Fig. 10.25 The antenna structure including the optimization parameters

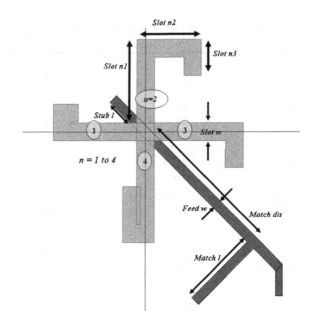

10.6.1 Example 1: Microstrip-fed Crooked Cross Slot Antenna

A microstrip-fed crooked cross slot antenna is optimized in this example. The antenna is a miniaturized form of the circular polarized crossed slot antenna based on work by Vlastis [25]. Compared to [25], each of the four arms of the radiating slots is broken clockwise into three parts to miniaturize the slot lengths and at the same time keep the CP performance. There are three length parameters (slot n1, slot n2, slot n3) on each arm of the slot structure shown in Fig. 10.25. We assume that all slot widths are equal. Therefore we have 13 variables for the four arms of the slot antenna. The advantage of this structure, shown in Fig. 10.25, is the simplicity of the single feeding structure. There is a stub added to the microstrip line to improve the matching (shown in pink). Therefore, there are four extra variable parameters on the feed line, one for the line width and three for the matching stub length, width and its position. These variables are shown in Fig. 10.25. The substrate used for this design is RT5880 1.5 mm and the center frequency for the design is 1.6 GHz.

The drawback of the structure is the unsymmetrical radiation pattern of the antenna due to its unsymmetrical structure, which rotates the radiation pattern and reduces the broadside gain. Therefore, the realized gain and the CP axial ratio (CPAR) are two important characteristics of the antenna together with the common requirement of return loss (S11). The antenna has about the same radiation on both the top and the bottom sides. Hence, the gain can be increased to almost twice when using an additional reflector at the bottom side. Therefore, the optimization problem is defined as:

Table 10.6 The optimized antenna (all values in mm)

Parameters	Optimized value	Lower bound	Upper bound
Slot 11	31.51	5	35
Slot 12	14.19	8	14
Slot 13	5.85	5	25
Slot 21	35	5	35
Slot 22	10.13	8	14
Slot 23	17.74	5	25
Slot 31	29.56	5	35
Slot 32	11.64	8	14
Slot 33	7.95	5	25
Slot 41	29.29	5	35
Slot 42	20.83	8	14
Slot 43	11.49	5	25
Slot w	5.64	2	6
Stub l	14.71	5	25
Match dis	50.86	30	55
Match l	25	5	25
Feed w	3.01	3	6.5

$$\text{minimize circular polarization axial ratio (CPAR)}$$
$$\text{s.t. realized gain} \geq 4\,\text{dB} \qquad\qquad (10.6)$$
$$S11 \leq -10\,\text{dB}$$

A penalty functions is used to integrate the two constraints into the objective function. In GPEME, the initial number of samples is set to 60 and all the other parameters are the same as those used in the benchmark problem tests in Sect. 10.4. The optimized design obtains 0.26 dB CPAR at 1.6 GHz, 4.34 dB realized gain and -13.4 dB S11 parameter. The optimized values of the design parameters are listed in Table 10.6 (all in millimeter).

The detailed performances simulated by ADS Momentum are as follows. From Fig. 10.26, it can be seen that the center frequency is at 1.575 GHz, and the bandwidth is 85 MHz (from 1.538 to 1.623 GHz). The CPAR versus θ angle is illustrated in Fig. 10.27. The antenna gain in terms of θ angle and its peak value (4.1 dB) at $\theta = -6°$ is illustrated in Fig. 10.28. It can be seen that the rotation of the beam is not significant and the broadside gain is almost 4 dB.

The total number of evaluations of GPEME is 268, costing 6 h. Then, the SBDE method is used. As was shown in Chap. 9, SBDE uses the standard DE with the tournament selection-based constraint handling method and without surrogate modeling. It costs about 1,200 iterations to converge, with a computational time of 28 h. The optimized performances are CPAR = 0.22 dB, realized gain = 4.53 dB, and S11 $= -13.56$ dB. It can be seen that the performance of the optimized antenna is only slightly better than that optimized by GPEME, but consuming about 5 times more computational effort. Note that the evaluation is quite cheap in this example (only

Fig. 10.26 The simulated
return loss of the antenna
structure optimized by
GPEME

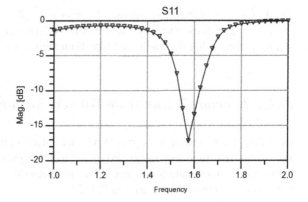

Fig. 10.27 Simulated AR
of the optimized antenna
versus θ

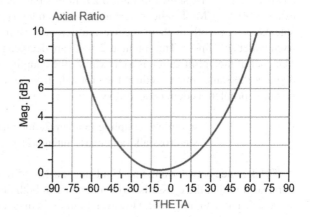

Fig. 10.28 Simulated
radiation pattern of the optimized antenna versus θ

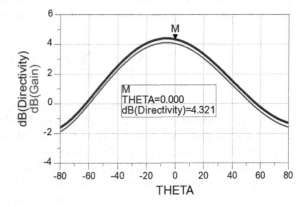

about 80 s per simulation) compared to other examples. If the EM simulation took from minutes to tens of minutes, which often occurs in complex antenna designs, the synthesis time using the traditional SBDE method would take from days to weeks.

10.6.2 Example 2: Inter-chip Wireless Antenna

Inter-chip antenna design is a recent popular antenna problem [26]. With the rapid growth in high-frequency integrated circuit technology, the new method of wireless inter-chip communication is proposed as an alternative solution with some advantages to wired chip interconnection [27, 28].

Example 2 analyzes a problem of short-range communication between three antennas at 60 GHz, shown in Fig. 10.29. In this inter-chip communication scheme using a 90 nm CMOS silicon technology, the antenna 2 communicates with the antenna 3, and both of them receive interferences from the antenna 1 which is a fixed wideband dipole. The antennas 2 and 3 are decided to be meander-line dipoles. The goal of the optimization is to maximize the coupling from antenna 2 to antenna 3 and at the same time to reduce the crosstalk from antenna 1. Therefore the optimization problem is defined as follows:

$$\begin{aligned} &\text{maximize coupling (antenna2, antenna3)}\\ &\text{s.t. crosstalk (antenna1, antenna2)} \le -30\,\text{dB} \end{aligned} \tag{10.7}$$

The distance between antenna 2 and antenna 3 is 2.5 mm (see Fig. 10.29). In order to make sure that the crosstalk from antenna 1 to both antenna 2 and 3 is the same, antenna 2 and 3 are mirrored. However, both of them can be asymmetrical as shown in Fig. 10.29. Each meander line antenna has 5 horizontal sections on each of the arms, namely $L1$ to $L5$ on one arm and $L6$ to $L10$ on the other arm. Thus, another constraint is the total length of each arm which is fixed to be 1 mm:

$$\sum_{m=1}^{5} L_m = \sum_{m=6}^{10} L_m = 1\,\text{mm} \tag{10.8}$$

This equality constraint can be handled in the algorithm's data generation.

The value of the antenna width La is 500 μm. The dipole length of antenna 1 is 2.6 mm and the dipole is situated at 1 mm from the other antennas.

In GPEME, the initial number of samples is set to 40 and all the other parameters are the same as those used in the benchmark problem tests in Sect. 10.4. The layouts of the optimized antennas are shown in Fig. 10.29, and the obtained values of the 10 design parameters are shown in Table 10.7. The coupling between antenna 2 and 3 is optimized to −18.25 dB while the constraint is satisfied with −30.18 dB crosstalk. Figure 10.30 shows the simulation details.

Table 10.7 Values of the variables of the optimized antenna 2 and 3 (μm) for antenna synthesis problem 2

L1	L2	L3	L4	L5	L6	L7	L8	L9	L10
8.34	136.80	8.34	823.33	23.19	328.85	328.85	328.85	6.76	6.69

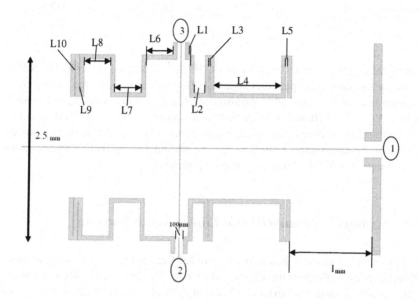

Fig. 10.29 Proximity coupling scheme and final layout of the optimized antennas (antenna problem 2)

Fig. 10.30 Simulated coupling from antenna 2 to 3 and the simulated crosstalk from antenna 1, optimized as a function of frequency at 60 GHz by GPEME

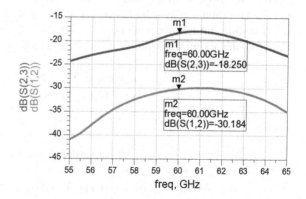

Table 10.8 Ranges of the 19 design variables (all sizes in mm)

Variables	A1	A2	A3	A4	A5	A6	B1	B2	B3	B4
Lower bound	4	4	4	12	4	4	6	4	4	20
Upper bound	12	12	12	30	12	28	20	16	8	40
Variables	B5	C1	C2	C3	C4	C5	C6	D1	X	
Lower bound	4	4	4	4	2	12	4	2	5	
Upper bound	16	12	12	12	26	30	12	24	12.5	

The total number of exact evaluations of GPEME is 302, taking 21 h. The SBDE method for the same example costs about 2,000 iterations to converge, with a computational time of 5.9 days. The optimized performances are coupling (antenna 2, antenna 3) = −18.84 dB and crosstalk (antenna 1, antenna 2) = −30.30 dB. It can be seen that the performance of the optimized antennas is comparable to that optimized by GPEME, but consuming 7 times more computational effort. This example shows the advantage of GPEME for more complex problems.

10.6.3 Example 3: Four-element Linear Array Antenna

A highly compact low-cost and strongly coupled four-element linear array antenna [29] is chosen as the third example in this subsection. This antenna has been optimized by the Particle Swarm Optimization (PSO) algorithm [24]. We therefore use PSO with the same settings in [24] but using the fitness function (10.9) as the comparison reference. The results obtained by the above reference method are better than that of [24]. The goal is to maximize the realized gain in the operating frequency range from 3.4 to 3.8 GHz. In this band, the S11 parameter should be less than −10 dB and the gain should at least be 13 dB. The substrate used is FR4. The performances at five equidistant frequency points (3.4, 3.5, 3.6, 3.7, and 3.8 GHz) are evaluated. Therefore the optimization problem is defined as follows:

$$\text{maximize } \sum_{i=1}^{5} RG_D^i$$
$$\text{s.t. } RG_D^i \geq 12.54 \text{ dB} \tag{10.9}$$

where

$$RG_D^i = gain(fr(i)) - 10 \times log10(1 - Reflection(fr(i))^2)$$
$$Reflection(fr(i)) = 10^{S11(fr(i))/20}$$
$$fr = \{3.4 \text{ GHz}, 3.5 \text{ GHz}, 3.6 \text{ GHz}, 3.7 \text{ GHz}, 3.8 \text{ GHz}\} \tag{10.10}$$

The topology of the four-element antenna is shown in Fig. 10.31 and the shape of the antenna is controlled by 19 design parameters indicated in the figure. The ranges of the design parameters are shown in Table 10.8.

A special character of this problem is that all the parameters must be integer numbers. However, the search engine of GPEME, the DE algorithm, aims at global optimization in a continuous space, instead of integer programming. Although truncation (quantization) methods can be used to apply DE to integer programming problems, it is still not widely used [30]. For integer programming and mixed integer-discrete optimization problems, methods based on swarm intelligence are shown to be a good choice [31], such as the ant colony algorithm (ACO). Therefore, to make DE workable for this example, two modifications are introduced to the standard GPEME. Firstly, the quantization method [17] is used. In all the DE search operators, floating numbers are always used, while these floating numbers are only rounded to the nearest integer in the function evaluation. Secondly, the mutation operator in DE is changed from DE/best/1 to DE/rand/1, where in the latter mutation operator three different vectors are randomly chosen, instead of using the current best candidate as the base vector [17]. The goal of these two modifications is to increase the diversity and the exploration ability as truncation makes the diversity decrease to some extent.

The EM simulator used is Magmas [21–23] and the evaluation of a candidate solution takes from 2 to 4.5 min depending on the size of the candidate design. The number of EM simulations is restricted to 600. 10 runs are carried out. Thus, the consumed computational time is from 10 to 14 h. In all the 10 runs using GPEME, the constraints are satisfied and the average objective function value is −71.05 dB. The variables of best result of the 10 runs are shown in Table 10.9 and the simulated performance is shown in Fig. 10.32. Compared to the reference method in [24], about 1,700 EM simulations are needed to obtain a comparable average objective function value as GPEME, so the speed enhancement of GPEME compared to PSO is nearly 3 times for this example. Using PSO, the average objective function value after 3,000 EM simulations is −71.65 dB. The convergence curve of GPEME is shown in Fig. 10.33. Note that this example is an integer programming problem, but the search engine of GPEME, the DE algorithm, is good at continuous optimization problems. On the other hand, the surrogate model-aware search framework in GPEME is also compatible with search engines (e.g., PSO, ACO) good at mixed integer-discrete optimization problems.

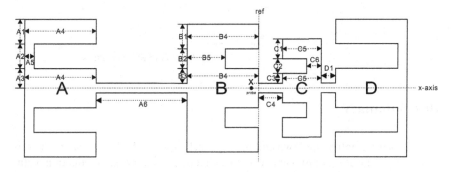

Fig. 10.31 Four-element antenna array top view (antenna example 3)

Table 10.9 The synthesized 19 design variables (best result) obtained by GPEME for antenna example 3

Variables	A1	A2	A3	A4	A5	A6	B1	B2	B3	B4
Values	12	12	12	30	4	28	20	4	4	28
Variables	B5	C1	C2	C3	C4	C5	C6	D1	X	
Values	4	12	4	4	26	30	12	24	8	

Fig. 10.32 Realized gain of the antenna synthesized by GPEME (best result) for antenna example 3

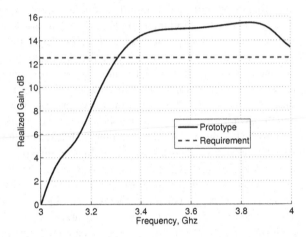

Fig. 10.33 GPEME convergence trend for the antenna example 3 in 600 EM simulations (average of 10 runs)

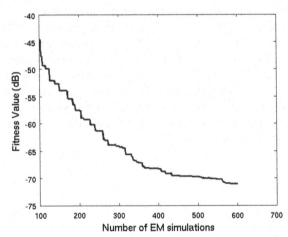

10.7 Summary

This chapter has introduced two cutting-edge techniques in SAEA research, the surrogate model-aware evolutionary search mechanism and dimension reduction methods based on Sammon mapping. The surrogate model-aware evolutionary search mechanism unifies the evaluation for optimization and the evaluation for surrogate modeling, so that the search can focus on a small promising area and is appropriately supported by the carefully constructed surrogate model. Using Sammon mapping to carry out dimension reduction on the search space for problems with around 50 variables, the model quality can be maintained with an affordable number of training samples and the computational overhead of modeling and prescreening can significantly be reduced. These are new directions to enhance the efficiency of SAEA for medium-scale expensive optimization problems. A method based on these techniques, GPEME, has been presented and was then applied to benchmark functions, a 60 GHz power amplifier synthesis and complex antenna synthesis, achieving high-quality designs in a practical time. GPEME is an efficient and effective design automation technique for general mm-wave integrated circuit design automation and complex antenna design automation.

This part of the book has introduced simulation-based electromagnetic design automation. In terms of SAEA research, important research issues include more effective dimension reduction methods, constraint handling in SAEA, especially in the surrogate model-aware evolutionary search framework, efficient SAEA for multi-objective expensive optimization, new SAEA frameworks for integrating parallel computation, etc. In a global picture, there are also possibilities such as integrating human-computer interaction into SAEA, especially multi-objective SAEAs. They will have a large impact both on the computational intelligence research and for the real-world applications, such as electromagnetic design automation.

Appendix

A. F1, F2, F3: Ellipsoid Problem

$$
\begin{aligned}
&min \quad f(x) = \sum_{i=1}^{d} i x_i^2 \\
&x \in [-5.12, 5.12], i = 1, \ldots, d \\
&F1 : d = 20, \ F2 : d = 30, \ F3 : d = 50 \\
&minimum : f(x^*) = 0
\end{aligned}
\tag{10.11}
$$

B. F4, F5, F6: Rosenbrock Problem

$$
\begin{aligned}
&min \quad f(x) = \sum_{i=1}^{d} (100(x_{i+1} - x_i^2)^2 + (1 - x_i)^2) \\
&x \in [-2.048, 2.048], i = 1, \ldots, d \\
&F4 : d = 20, \ F5 : d = 30, \ F6 : d = 50 \\
&minimum : f(x^*) = 0
\end{aligned}
\tag{10.12}
$$

C. F7, F8, F9: Ackley Problem

$$min \quad f(x) = -20e^{-0.2\sqrt{\frac{1}{d}\sum_{i=1}^{d}x_i^2}} - e^{\frac{1}{d}\sum_{i=1}^{d}cos(2\pi x_i)}$$
$$x \in [-32.768, 32.768], i = 1, \ldots, d$$
$$F7 : d = 20, F8 : d = 30, F9 : d = 50 \tag{10.13}$$
$$minimum : f(x^*) = 0$$

D. F10, F11, F12: Griewank Problem

$$min \quad f(x) = 1 + \sum_{i=1}^{d}\frac{x_i^2}{4000} - \prod_{i=1}^{d}cos(\frac{x_i}{\sqrt{i}})$$
$$x \in [-600, 600], i = 1, \ldots, d$$
$$F10 : d = 20, F11 : d = 30, F12 : d = 50 \tag{10.14}$$
$$minimum : f(x^*) = 0$$

E. F13, Shifted Rotated Rastrigin Problem

$$min \quad f(x) = \sum_{i=1}^{d}(z_i^2 - 10cos(2\pi z_i) + 10) - 330$$
$$z = (x - o) * M$$
$$x \in [-5, 5], i = 1, \ldots, d \tag{10.15}$$
$$F13 : d = 30$$
$$minimum : f(x^*) = -330$$

The values of M and o can be found in [16].

F. F14, Rotated Hybrid Composition Function with Narrow Basin Global Optimum
Its details can be found in [16].

References

1. Liu B, Zhang Q, Gielen G (2013b) A gaussian process surrogate model assisted evolutionary algorithm for medium scale expensive black box optimization problems. IEEE Trans Evol Comput (To be published)
2. Lim D, Jin Y, Ong Y, Sendhoff B (2010) Generalizing surrogate-assisted evolutionary computation. IEEE Trans Evol Comput 14(3):329–355
3. Emmerich M, Giannakoglou K, Naujoks B (2006) Single-and multiobjective evolutionary optimization assisted by gaussian random field metamodels. IEEE Trans Evol Comput 10(4):421–439
4. Zhou Z, Ong Y, Nair P, Keane A, Lum K (2007) Combining global and local surrogate models to accelerate evolutionary optimization. IEEE Trans Syst Man Cybern Part C Appl Rev 37(1):66–76
5. Jones D (2001) A taxonomy of global optimization methods based on response surfaces. J Global Optim 21(4):345–383
6. Jin Y (2005) A comprehensive survey of fitness approximation in evolutionary computation. Soft Comput Fusion Found Methodologies Appl 9(1):3–12
7. Gorissen D, Couckuyt I, Demeester P, Dhaene T, Crombecq K (2010) A surrogate modeling and adaptive sampling toolbox for computer based design. J Mach Learn Res 11:2051–2055
8. Van Der Maaten L, Postma E, Van Den Herik H (2008) Dimensionality reduction: a comparative review. Published online 71(January), pp 2596–2603

9. Sammon J Jr (1969) A nonlinear mapping for data structure analysis. IEEE Trans Comput 100(5):401–409
10. Henderson P (2007) Sammon mapping. Published online, http://wwwhomepagesinfedacuk pp 1–5
11. Avriel M (2003) Nonlinear programming: analysis and methods. Dover Publishers, New York, USA
12. Stein M (1987) Large sample properties of simulations using Latin hypercube sampling. Technometrics 29(2):143–151
13. Jones D, Schonlau M, Welch W (1998) Efficient global optimization of expensive black-box functions. J Global Optim 13(4):455–492
14. Zhang Q, Liu W, Tsang E, Virginas B (2010) Expensive multiobjective optimization by MOEA/D with gaussian process model. IEEE Trans Evol Comput 14(3):456–474
15. Eiben A, Bäck T (1997) Empirical investigation of multiparent recombination operators in evolution strategies. Evol Comput 5(3):347–365
16. Suganthan P, Hansen N, Liang J, Deb K, Chen Y, Auger A, Tiwari S (2005) Problem definitions and evaluation criteria for the CEC 2005 special session on real-parameter optimization. Nanyang Technological University, Singapore, Technical Report 2005005
17. Price K, Storn R, Lampinen J (2005) Differential evolution: a practical approach to global optimization. Springer-Verlag, New York
18. IEEE (2010) WiGig, MAC and specification, PHY v1. 0
19. IEEE 80215 Working Group (2009) Wireless PAN task group 3c. millimeter wave alternative PHY.http://www.ieee802.org/15/pub/TG3c.html
20. Agilent (2013) Agilent Technology homepage. http://www.home.agilent.com/
21. Vandenbosch G, Van de Capelle A (1992) Mixed-potential integral expression formulation of the electric field in a stratified dielectric medium-application to the case of a probe current source. IEEE Trans Antennas Propag 40(7):806–817
22. Demuynck F, Vandenbosch G, Van de Capelle A (1998) The expansion wave concept. I. efficient calculation of spatial green's functions in a stratified dielectric medium. IEEE Trans Antennas Propag 46(3):397–406
23. Schols Y, Vandenbosch G (2007) Separation of horizontal and vertical dependencies in a surface/volume integral equation approach to model quasi 3-D structures in multilayered media. IEEE Trans Antennas Propag 55(4):1086–1094
24. Ma Z, Vandenbosch G (2012) Comparison of weighted sum fitness functions for PSO optimization of wideband medium-gain antennas. Radioengineering 21(1):504–511
25. Vlasits T, Korolkiewicz E, Sambell A, Robinson B (1996) Performance of a cross-aperture coupled single feed circularly polarised patch antenna. Electron Lett 32(7):612–613
26. Lin J, Wu H, Su Y, Gao L, Sugavanam A, Brewer J et al (2007) Communication using antennas fabricated in silicon integrated circuits. IEEE J Solid-State Circuits 42(8):1678–1687
27. Kim K, Floyd B, Mehta J, Yoon H, Hung C, Bravo D, Dickson T, Guo X, Li R, Trichy N et al (2005) On-chip antennas in silicon ICs and their application. IEEE Trans Electron Devices 52(7):1312–1323
28. Drost R, Hopkins R, Ho R, Sutherland I (2004) Proximity communication. IEEE J Solid-State Circuits 39(9):1529–1535
29. Volski V, Delmotte P, Vandenbosch G (2004) Compact low-cost 4 elements microstrip antenna array for WLAN. In: Proceedings of 7th european conference on wireless technology, pp 277–280
30. Gao J, Li H, Jiao Y (2009) Modified differential evolution for the integer programming problems. In: Proceedings of international conference on artificial intelligence and computational intelligence, vol 1. pp 213–219
31. Bonabeau E, Dorigo M, Theraulaz G (1999) Swarm intelligence: from natural to artificial systems. Oxford University Press, New York, USA

Printed in the United States
By Bookmasters